PALGRAVE STUDIES IN CULTURAL AND INTELLECTUAL HISTORY
Series Editors
Anthony J. La Vopa, North
Suzanne Marchand, Louisiana
Javed Majeed, Queen Mary, University

The Palgrave Studies in Cultural and Intellectual History series has three primary aims: to close divides between intellectual and cultural approaches, thus bringing them into mutually enriching interactions; to encourage interdisciplinarity in intellectual and cultural history; and to globalize the field, both in geographical scope and in subjects and methods. This series is open to work on a range of modes of intellectual inquiry, including social theory and the social sciences; the natural sciences; economic thought; literature; religion; gender and sexuality; philosophy; political and legal thought; psychology; and music and the arts. It encompasses not just North America but Africa, Asia, Eurasia, Europe, Latin America, and the Middle East. It includes both nationally focused studies and studies of intellectual and cultural exchanges between different nations and regions of the world, and encompasses research monographs, synthetic studies, edited collections, and broad works of reinterpretation. Regardless of methodology or geography, all books in the series are historical in the fundamental sense of undertaking rigorous contextual analysis.

Published by Palgrave Macmillan:

Indian Mobilities in the West, 1900–1947: Gender, Performance, Embodiment
By Shompa Lahiri

The Shelley-Byron Circle and the Idea of Europe
By Paul Stock

Culture and Hegemony in the Colonial Middle East
By Yaseen Noorani

Recovering Bishop Berkeley: Virtue and Society in the Anglo-Irish Context
By Scott Breuninger

The Reading of Russian Literature in China: A Moral Example and Manual of Practice
By Mark Gamsa

Rammohun Roy and the Making of Victorian Britain
By Lynn Zastoupil

Carl Gustav Jung: Avant-Garde Conservative
By Jay Sherry

Law and Politics in British Colonial Thought: Transpositions of Empire
Edited by Shaunnagh Dorsett and Ian Hunter

Sir John Malcolm and the Creation of British India
By Jack Harrington

The American Bourgeoisie: Distinction and Identity in the Nineteenth Century
Edited by Sven Beckert and Julia B. Rosenbaum

Benjamin Constant and the Birth of French Liberalism
By K. Steven Vincent

The Emergence of Russian Liberalism: Alexander Kunitsyn in Context, 1783–1840
By Julia Berest

The Gospel of Beauty in the Progressive Era: Reforming American Verse and Values
By Lisa Szefel

Knowledge Production, Pedagogy, and Institutions in Colonial India
Edited by Indra Sengupta and Daud Ali

Religious Transactions in Colonial South India: Language, Translation, and the Making of Protestant Identity
By Hephzibah Israel

Cultural History of the British Census: Envisioning the Multitude in the Nineteenth Century
By Kathrin Levitan

Character, Self, and Sociability in the Scottish Enlightenment
Edited by Thomas Ahnert and Susan Manning

The European Antarctic: Science and Strategy in Scandinavia and the British Empire
By Peder Roberts

Origins of Modern Historiography in India: Antiquarianism and Philology, 1780–1880
By Rama Sundari Mantena

Isaiah Berlin: The Journey of a Jewish Liberal
By Arie Dubnov

Making British Indian Fictions: 1772–182
By Ashok Malhotra

Alfred Weber and the Crisis of Culture, 1890–1933
By Colin Loader

Alfred Weber and the Crisis of Culture, 1890–1933

Colin Loader

palgrave
macmillan

ALFRED WEBER AND THE CRISIS OF CULTURE, 1890–1933
Copyright © Colin Loader, 2012.

Softcover reprint of the hardcover 1st edition 2012 978-1-137-03114-3
All rights reserved.

First published in 2012 by
PALGRAVE MACMILLAN®
in the United States—a division of St. Martin's Press LLC,
175 Fifth Avenue, New York, NY 10010.

Where this book is distributed in the UK, Europe and the rest of the world, this is by Palgrave Macmillan, a division of Macmillan Publishers Limited, registered in England, company number 785998, of Houndmills, Basingstoke, Hampshire RG21 6XS.

Palgrave Macmillan is the global academic imprint of the above companies and has companies and representatives throughout the world.

Palgrave® and Macmillan® are registered trademarks in the United States, the United Kingdom, Europe and other countries.

ISBN 978-1-349-44074-0   ISBN 978-1-137-03115-0 (eBook)
DOI 10.1057/9781137031150

Library of Congress Cataloging-in-Publication Data is available from the Library of Congress.

A catalogue record of the book is available from the British Library.

Design by Newgen Imaging Systems (P) Ltd., Chennai, India.

First edition: August 2012

For
Michael and Claire

Contents

Acknowledgments	ix
Introduction	1

Part I Alfred Weber in the German Empire, 1890–1918

1	The Context of Alfred Weber's Early Work	13
2	Early Economic Writings, 1897–1908	31
3	Heidelberg and the Empire, 1907–1917	47
4	The Question and Sociology of Culture	61
5	The Cultural Theory of Politics	81

Part II Alfred Weber in the Weimar Republic

6	The Weimar Republic and the End of the Discursive Coalition	99
7	The Sociology of Culture in Weimar	115
8	Cultural Politics in Weimar	133
9	Epilogue: Alfred Weber After 1933	157

Appendix 1: A Note on the Translation	163
Appendix 2: Fundamentals of Cultural Sociology: Social Process, Civilizational Process and Cultural Movement (1920)	165
Appendix 3: Cited Works of Alfred Weber	205
Notes	211
Index	243

Acknowledgments

A number of people helped this book come to fruition. The late Friedrich Tenbruck encouraged me to pursue the project. A sabbatical leave from the University of Nevada, Las Vegas, and a research stipendium from the Deutsche Akademische Austauschdienst in the early 1990s gave me the opportunity to spend a year researching at the University of Trier. Hans Braun sponsored my stay in the Faculty for Economic and Social Sciences there, and has continued to offer friendship and assistance with my return to UNLV. At Trier, I was able to investigate the rich institutional literature on the social sciences during the empire and republic by German scholars, which has largely been ignored in this country. At that same time, there was a renewed interest in Weber in Germany, led by Eberhard Demm. I was able to benefit from that, and Demm in particular has been very helpful to me in my studies, in addition to reading a draft of the manuscript. David Kettler, while luring me away from Weber to other projects, has been an inspiring collaborator and engaging discussion partner over the years. Parts of the manuscript were presented in the UNLV History Department's faculty seminar, where I received helpful criticism, especially from Paul Werth, Michelle Tusan, Kelly Mays, and Mary Wammack. Mary and I have discussed our similar methodological approach despite our divergent interests, over the years and, as she has done for so many others in our department, she proofread the entire manuscript. Anonymous readers for the press have offered useful suggestions that have improved the book. Editors Anthony La Vopa and Chris Chappell provided the necessary support and guidance; they made working with Palgrave Macmillan a pleasure. This is not an all-inclusive list; there are others not mentioned here who provided assistance, encouragement, and sometimes just a listening ear. I thank them all. All faults in the book are entirely my own.

Introduction

Most sociological theorists would view the years from 1890 to 1930 as the seminal period of their discipline. There were individual great social thinkers such as Karl Marx who preceded that era, but they were not part of a "cluster of genius"[1] matching the collective brilliance of their successors. Germany, where sociology emerged as a significant intellectual (if not institutional) force during the period, made an important contribution to this cluster. The voluminous literature on Max Weber and Georg Simmel gives testimony to their inclusion in almost every history of the human sciences. Many would add Karl Mannheim as a junior partner of the duo. Although these thinkers often have been treated abstractly in isolation as theorists (with varying results), most interpreters understand that they were part of their own historical ecosystem consisting of political, social, economic, and cultural institutions, as well as a second tier of "sociologists" who tackled the same issues.[2] While some might deny the importance of this context in understanding the work of its greatest participants, I concur with Wilhelm Hennis, who, in 1983, wrote about Max Weber:

> Not only does the socio-cultural background of his generation await study, but the *Weltanschauung* and the scientific problems in which Weber's generation so passionately engaged are even further removed from us. We know far too little about these factors, or at any rate not enough for an understanding of Weber's work in this context.[3]

Since (and even before[4]) Hennis penned his lament, steps have been taken to remove this lacuna. This book can be viewed as one of those steps.

It examines the intellectual career of Alfred Weber, Max's brother and Mannheim's sponsor. Max Weber and Simmel were somewhat marginal institutional figures within academia, the former due to his debilitating illness, the latter due to systemic anti-Semitism, as was the academic discipline of sociology during their lifetimes. Neither lived to witness sociology's partial institutionalization in the succeeding Weimar Republic. Alfred Weber, in contrast, did. He played a more important role within the university and engaged intellectually with members of the next two generations of sociologists—including Mannheim, Norbert Elias, and Hans Gerth—as well as with intellectuals outside of academia.

Wolfgang Schluchter is correct that Alfred Weber never approached the legacy of his older brother,[5] but Alfred was a pioneer in the establishment of two areas of investigation that have come to be associated with Max—cultural sociology and the critique of bureaucracy. His formulations of cultural sociology, one of which is included as an appendix, were among the first to attempt to define that subdiscipline. It has recently been argued that Alfred's study of the cultural impact of the German hyperinflation on intellectuals "echoes the cultural sociology of Pierre Bourdieu,"[6] a claim I will address in Chapter 8 below. His important address on officialdom (bureaucracy) not only preceded the famous study by Robert Michels, who cited it the next year,[7] but also inspired Franz Kafka's novel *The Trial*. Weber's seminar in Heidelberg was a center of cultural sociology in the 1920s. That was where Mannheim began his path to the sociology of knowledge. Mannheim's answers to the cultural crisis of Weimar might have received more attention than those of his mentor, but the questions raised were those of Weber. Whatever their staying power, Weber's works retain their importance as founding documents of the discipline.

Weber's role as a political actor has been thoroughly investigated by Eberhard Demm, who led a resurgence of interest in Weber in Germany during the last three decades.[8] Nevertheless, he is largely a forgotten figure in the United States, despite the enormous popularity of his brother Max.[9] This book will attempt to remedy that neglect. It will investigate not only his cultural sociology and (briefly) the succeeding cultural-historical studies, but also the very different economic writings that preceded it and the cultural-political program that accompanied it. The emphasis will be historical, to place Weber in his German context and to view his cultural sociology as an attempt to diagnose and deal with the problems of that context. In a sense this study is as much about that context as it is about Weber.[10]

In focusing on the interaction of an individual with his larger context, I have tried to negotiate between biography, the history of ideas, and institutional studies. I have made use of all three of those aspects, but the emphasis is on the interaction among them. I have mostly avoided tracing individual influences on Weber in favor of examinations of how he approached questions that grew out of more general contextual issues in comparison to the positions of other individuals. This strategy confronts the same problem experienced by thinkers at the time, namely, how to classify and organize the complex social field in which these thinkers wrote. My approach has always been a modified version of the historical sociology of knowledge formulated by Alfred Weber's onetime protégé, Karl Mannheim.[11] Like Mannheim, I am heedful of Max Weber's warning that theory be used as a heuristic tool in historical research and not as a set of generalizations to be proven by such research. Accordingly, I have not hesitated to significantly modify Mannheim's approach by borrowing from other theories, which I have in turn simplified and modified. My purpose is not to create a better theory, but to use a variety of concepts to better understand the interaction of the man and his context.

In discussing Alfred Weber's immediate context during his formative years, I have borrowed Peter Wagner's concept of "discursive coalition,"[12] which borrows from the theory of Pierre Bourdieu.[13] The latter discusses how individuals and institutions create self-referential systems of meaning that reduce complexity and thereby establish meaning and normative values for their participants. Individuals, through their interaction and communication, create institutional structures and systems of meaning, "discourses," which in conjunction with one another shape the understandings and influence the behavior of individuals. To describe this reciprocality, Bourdieu uses the concept of "habitus," which he defines as

> systems of durable, transposable dispositions, structured structures predisposed to function as structuring structures, that is, as principles which generate and organize practices and representations that can be objectively adapted to their outcomes without presupposing a conscious aiming at ends or an express mastery of the operations necessary in order to attain them. Objectively "regulated" and "regular" without being in any way the product of obedience to rules, they can be collectively orchestrated without being the product of the organizing action of a conductor.[14]

Discourse and habitus are connected to and help define relationships of power, wealth, and status among individuals and institutions. Because the ability to read and interpret discourse is unevenly distributed, those privileged with access to the appropriate cultural tools for interpreting the discourse attain elevated social status, or "symbolic power." This power, which is built upon the "cultural capital" of privileged knowledge, and which is convertible into other forms of capital—political, social, and/or economic—perpetuates itself within the system. But the flow of cultural capital to and from other forms of capital is always incomplete and often results in tensions between their possessors.[15] None of the elements that comprise the field are homogeneous or immutable. They are always sites of contestation in which some forms are more successful than others. And the field itself undergoes modification as its components do. This is especially true in times of perceived crisis, when assumptions of the habitus are increasingly questioned. When this happens, the relationship between the possessors of symbolic power and other members of society is often altered. Such changes occurred during the years 1897–1933, which can certainly be seen as times of crisis.[16] They contained the transformative force of rapid industrialization, the resultant growth of an active socialist movement, a world war, major economic dislocations of inflation and depression, political turmoil including the emergence of fascism, and an uncertain place in international diplomacy. Alfred Weber's work was designed to provide orientation during these turbulent times.

I examine Weber's attempts at orientation by examining the field he attempted to diagnose: how certain elements of it were engrained in him; how he modified those and contested other elements; if and how other individuals in the field reacted to his interventions and vice versa. Given the complexity of any field, an attempt to examine its interaction with an individual within it will contain an element of distortion in its presentation. Rather than attempt to offer a complete or balanced view of the entire field, I focus on those lines of force in closest proximity to Weber.

I start with Niklas Luhmann's description of the systemic changes that occurred in nineteenth-century Germany, where a purely stratificatory system gave way to a more functional one. He identifies three major types of functional differentiation (economic, political, and educational) that interacted and potentially competed with one another. The German national system's self-definition required that this competition among its subsystems be resolved.[17] The economic system had its unity in money, the medium in which all economically relevant relations and operations were precisely expressed. The political and educational systems did not possess the same technical precision as the economic and thus required additional self-description to achieve the same integrative capacity. The political system used the concept of the "state" to provide the unifying function for "meaning reference of all operations that claim to function as elements in the political system." The state provided the same kind of self-reference in the political sphere that money did in the economy. The educational system relied on the concept of "cultivation" (*Bildung*) to play this self-referential role. The ideal of individual learning "now became an individualized correlate of the world."[18] Luhmann writes that German elites were unwilling to accept three separate systems of self-reference.

> If one looks back on the semantic careers of the self-descriptive concepts like [economic] "capital," "state," and *Bildung*, it becomes apparent that, especially in the German academic tradition, attempts were repeatedly undertaken, not to come to terms with difference, but to integrate it in the name of a holistic formula.[19]

This holistic strategy allied the political and educational systems in a symbiotic relationship to subordinate the more universalistic economic system. The holders of political and educational capital zealously sought to ensure that those forms were clearly distinguished from and superior to economic capital. This coalition can be described as "discursive," because its articulation contains what I see as the two primary elements that comprise a discourse, conceptual contents and a structural organization of those concepts.[20]

The organization of the study

The book is divided into two parts. The five chapters of Part I contain an examination of the imperial discursive coalition, Weber's place within it as a young scholar, the importance of the setting of Heidelberg on his work after 1907, and his oppositional stance to the coalition's political and cultural priorities despite the continuance of some of its structural assumptions in his writings.

Chapter 1 examines the coalition and especially the discipline that was located at its academic center in the late nineteenth century, the Historical School of National Economy,[21] within which Weber was trained. His mentor, Gustav Schmoller, emphasized the importance of the state and its officials, most of whose qualifications rested on a certain level of academic learning (*Bildung*), in providing order and values for the nation. Like others in his generation, Weber came to question the older generation's emphasis on the state and its organic qualities. But while they challenged the specific concepts of the coalition, its binary structure continued in modified form in their writings.

Chapter 2 examines the two sets of Weber's economic writings. His earliest studies—specialized empirical investigations of the clothing industry—subscribed to the Historical School's belief in the importance of the state as a social actor. In 1904 he ventured outside of the specific German sphere by accepting a professorship in Prague within the Austro-Hungarian Empire. There he temporarily adopted an abstract approach that was more akin to the rival Austrian School of Marginal Utility than to the Historical School. This change resulted in an important book, *Theory of the Location of Industry*.

Chapter 3 examines the setting of Heidelberg, where Weber was called to become a full professor in 1907. Although he had returned to Germany, Weber continued to be distanced from the orthodoxy of the Historical School in the half-decade before World War I, but in a completely different direction from that of abstract economic theory. His adoption of a monadic organic theory of culture, which allowed for individual uniqueness and heterogeneity as characteristics of an organic unity, was compatible with the intellectual environment of the university town, as was his place in a generational revolt (that included his brother) against the discursive coalition with its center in the capital city of Berlin. His opposition consisted of two basic elements: a theory of culture that replaced the coalition's emphasis on the state; and a new approach to the political sphere based on his cultural theory.

Chapter 4 examines Weber's views on culture prior to World War I. Heavily influenced by vitalism (*Lebensphilosophie*), which had become increasingly popular in Germany after 1890, he sought a new cultural configuration that would allow for a diversity of expression while maintaining a coherent meaning. He wrote that the center for cultural renewal—actually centers—had to develop within individuals and that there would be some sort of cultural harmony among the centers because they all participated in life through feeling. During this same period, Weber tried to establish a new discipline, cultural sociology, which he organized upon the dualism of "civilization" and "culture," to promote his agenda.

He described the civilizational process as the intellectual and technical realization of the basic needs that constituted the social process. It was essentially utilitarian and instrumentally rational. It was also universal and was characterized by the Enlightenment idea of progress. Individual historical entities could either hinder or accelerate its development, but the process itself superseded them. One could say that this process was akin to the rationalization process that his brother Max saw as most advanced in the Western world and that is often identified with "modernity." Culture was very different, consisting of an aggregation of symbols for the purpose of making sense of the existing configuration of the world. It was especially active when there were "reaggregations" of the "life-elements." It was neither universal nor unilinear, but was connected and confined to specific historical entities. It did not possess the rational continuity of the civilizational process, but rather was discontinuous and thus historically relativistic. It could also be described as organic, as opposed to the mechanism of the civilizational process.

In its basic form this dualism was not at odds with the mainstream academic configuration, which was also binary, but the elements organized by this pattern were rearranged. The most important of these elements was the state, which held a central place in the view of the academic establishment, one closely connected with organic cultural values. Weber, following Ferdinand Tönnies, moved the state to the civilizational side of the dualism, thus creating a gap between it and the creation of cultural values. But he did not follow Tönnies's pessimistic temporal pattern: civilization interacted with culture rather than succeeding it. This approach, Weber believed, opened the way for a new relationship between the creators of cultural values and political institutions. He dedicated his career and the discipline of cultural sociology to the definition of such a relationship. He charged cultural sociology with two tasks: the investigation of the relationship between concrete individual creations and the organic cultural center, and the relating of both of these elements to the civilizational process and its objectifications. In doing so, cultural sociology, said Weber, moved beyond sociology proper. Although Weber's formulation of his discipline was properly criticized for its vagueness,[22] it remained politically engaged. He believed that the inability of people to integrate developments in the civilizational process into any coherent unity of values resulted in cultural crisis and political instability. He hoped the discipline would bridge what he saw as an increasing gap between academia and a larger public, especially with regard to the political process.

Chapter 5 discusses Weber's application of his cultural sociology to the political sphere. He challenged the coalition's notion that the state established an organic realm of meaning that transcended the divisive economic sphere. In an important prewar essay titled "The Official" (1910),[23] he accused the political establishment and its academic supporters of undermining their own cultural claims by promoting a mechanistic bureaucratic structure. He redefined the "Social Policy" of the Historical School—the integration of the rebellious working classes into the national community politically and economically—as a "crisis of culture." In fact, he wrote, the established Social Policy simply attempted to extend to the working classes the same process of bureaucratization that had already captured the upper and middle classes. It took workers out of their previously "free existence" and turned them into mere labor power, thereby throwing them into "that gray, barren, uniform housing (*Gehäuse*)" of bureaucratized existence. Weber applied his organic theory of culture to two broad areas of politics: domestic leadership and its relationship to the lower classes and foreign policy. His approach to both areas was determined by his views on culture. Prior to World War I, his primary concern was domestic issues, especially the impact of the dramatic societal change and the challenges it posed. During the war, he emphasized foreign policy over domestic policy, but as the war concluded and the prospects for democracy increased, the emphasis was again reversed.

Part II covers Weber's career after the dissolution of the coalition in 1918, including his new Weimar context and the lingering effects that the old discursive habitus had on his ideas. When Germany's defeat in World War I and the subsequent revolution brought an end to the ruling coalition, Weber's positions changed accordingly. He participated in groups and institutions that he felt would promote his vision of a democratic society informed by cultural concerns. Chapter 6 briefly looks at the Weimar Republic and its impact on Heidelberg and Weber's career.

Chapter 7 examines Weber's expansion of his writings on cultural sociology. In doing so, he faced the challenge of new intellectual competitors. The two most important were his deceased brother Max and the most popular vitalist of the era, Oswald Spengler. Weber's most important writings on the discipline (one of which appears here in translation as an appendix) come from this period. In 1920, he settled on culture as a "movement" to distinguish it from the "process" of civilization, and he added a third term, "social process" (which had made a brief appearance in his lectures in 1910), but it never achieved the importance of the first two. In addition to expanding on the nature of these three spheres, he elaborated on the methodology of the discipline, separating his approach from that of his brother Max. His two most important applications of cultural sociology, his account of the distress of German intellectuals and his short cultural history of ancient Egypt and Babylonia, appeared in the mid-1920s. As the republic collapsed, Weber's cultural sociology took an unfortunate turn, becoming a metahistory in the manner of Spengler. Fellow sociologist Leopold von Wiese accused him of having left rigorous sociology for "dilettantism."[24] A series of books, beginning during the final years of Weimar with *Cultural History as Cultural Sociology* and resumed with the end of the war, were designed to answer the question: where do we find ourselves in the stream of history?[25] This series of works, which Weber himself held as his most significant, in fact only supports Wiese's judgment.

Chapter 8 presents Weber's Weimar writings on domestic politics and foreign affairs, which followed a similar pattern to those before the republic. Central to his concerns was the question of spiritual (*geistige*) leadership, whether it be a new spiritual elite to lead the democratic masses within Germany or Germany's assumption of spiritual leadership for the smaller nationalities of Central Europe. Weber believed that this spiritual leadership within Germany would emerge from the youthful elements of the various democratic institutions and assume a spiritual-aristocratic posture. The latter would separate personal standards from party opinion and could work out a spiritual-aristocratic norm. For Weber, it was important to understand that a spiritual aristocracy was not incompatible with modern democracy. He termed such an arrangement "leader democracy." In foreign policy, he wanted the individuality of the peoples of Central Europe to be preserved with the assurance that this would promote their organic unity. Two principles, political multiplicity and spiritual unity, the latter characterized by the belief in general human rights, remained at the core of this cultural aspiration, which was to be realized in some type of federalism. And he further optimistically assumed that Germany would provide the necessary leadership.

During the last years of the republic, Weber became increasingly pessimistic about whether a new cultural synthesis could be created domestically or in foreign relations. Although he was a vigorous critic of fascism, he admitted that the Nazis were better equipped to deal with the new reality because they had developed a new vision and a new "*élan.*" He hoped a new democratic leadership detached from the parliamentary process could contest the fascist vision. This leadership, or any other capable of saving Weimar democracy, never emerged, and, with the decline of the republic, Weber largely retreated from active participation in parliamentary politics.

A brief epilogue surveys Weber's emergence from his "internal emigration" under Nazism to play a significant role in the postwar Allied occupation and the early Federal Republic, especially in the attempt to denazify Heidelberg University. I believe that his cultural writings during this later period add little to his earlier work.[26] His political involvement has been well examined in detail by Demm and others.

There has been much written on the relationship of culture and politics in Germany. This literature has often taken an either/or form: the obsession of German elites, especially academic elites, with culture resulted in a rejection of parliamentary politics. The title of the latest entry in this series, Wolf Lepennies's *The Seduction of Culture in German History*, speaks for the genre when it declares that "culture was seen as a noble substitute for politics."[27] Yet, throughout his career, Weber, like his brother Max, was dedicated to both culture and democratic politics. The problem with his approach, unlike that of his brother, lay in his specific monadic concept of culture. Although there were changes in Weber's cultural writings, as well as changes in the settings in which they appeared, throughout his career they were premised on a pedagogical mission to create a new individuality in both the elites and the democratic masses. This individuality would be monadically tied to a larger organic unity. Weber's inability to abandon this organic individual-whole relationship resulted in the attempt to wed a specific type of culture to democratic politics. Weber's mission to realize this relationship began as a response to the imperial context of issues, and even with the demise of the empire it remained informed by its original habitus. It is to that setting that we now turn.

Part I
Alfred Weber in the German Empire, 1890–1918

1
The Context of Alfred Weber's Early Work

Alfred Weber's career began in the last decade of the nineteenth century, a period of dramatic change in Germany. In a brief autobiographical sketch he described his family background, which has been well documented in studies of his brother Max.[1] He emphasized the sense of political and intellectual engagement that his father, a National Liberal politician, and his brothers imparted to him.

> Political interests and, as far as could be accomplished, political activity went without saying for him and his brother (Max Weber, died 1920). However, both realized that, since the disempowerment of parliament after 1878, a political career was meaningless given the current dominant pseudo-constitutionalism. They restricted themselves, therefore, to public criticism in the press, to speeches and constructively performed activity in attempts at economic and social reform.[2]

This passage points to a tension between the authoritarian governments of Prussia and the empire and those forces that advocated parliamentarization and reform. The lower house of the imperial parliament, the *Reichstag*, was "weak" despite the democratic veneer of universal manhood suffrage. It was limited to the "negative politics" of rejecting legislation and bore no responsibility for constructing a government, which remained the prerogative of the emperor. This "pseudo-constitutional" arrangement was aligned with an array of allies, the Prussian *Junker*, the army and upper bureaucracy, and important elements within the industrial bourgeoisie. This chapter focuses on the one element of this coalition that articulated its legitimizing discourse, the academic establishment.

This coalition was not free of opposition, some of it extending into the very institution that articulated its defense, the university. Industrial growth in the last quarter of the century brought with it significant social and demographic changes. Although it fostered the "alliance of rye and steel," it also gave rise to voices challenging the power structure, most notably the rapidly growing Social Democratic Party with its demand for more parliamentary responsibility. Under the new Kaiser, the mercurial Wilhelm II, Germany embarked on a "new course" after 1890. The new administration attempted to address what was termed the "social question," the integration of the working classes into the national polity. A series of social reform decrees were passed that proclaimed the attempt to arrive at a "social peace" in Germany. These reforms had a significant impact on a generation of Germans who came of age in the 1890s, one of whom was Alfred Weber. The result was a "reform milieu" consisting of a wide spectrum of groups both inside and outside the establishment. Other forces in addition to proletarian socialism overlapped with this reform milieu, placing increased demands on the establishment. A rural populism with disturbing irrational traits such as anti-Semitism and antimodernism, which found additional supporters among the upper classes, also emerged. A disparate counterculture appeared, including a youth movement, a new feminism and eroticism, and enthrallment with vitalism and the cultural nihilist Friedrich Nietzsche. These forces formed, in Kevin Repp's words, a "matrix of shared issues and objectives."[3]

> In one way or another, each member of the milieu interacted with this matrix, which included concerns for the impact of capitalism and industrialization, the purposive application of modern science and technology, the need for social order and efficiency, the creation of an integral national community, the maintenance of humanistic values, and a predilection for organicist models of development and synthesis. This is not to say that each member shared an interest in all of these issues.... [O]nce its structures had gained sufficient currency, [this matrix] established its own boundaries on the complex map of divided loyalties in Imperial political culture.[4]

As Repp indicates, this "matrix" and the larger "social field" in which it was embedded did not consist of unambiguously demarcated groups and institutions. "Entities" such as the state, society, and modernity were not really entities but rather complexes of interacting and overlapping heterogeneous elements. For example, some "state" functions were carried out by elements that might be better described as belonging to society, thus blurring the distinction between the two. And "modern" and "antimodern" elements were alternatively opposed to and affiliated with one another.[5]

Alfred Weber entered within this milieu in the late 1890s, when he became part of the discipline of national economy within the university system. Establishment national economists attempted to mediate among the conflicting elements in the social field, endeavoring to reconcile them with the ruling coalition. As the defenders of the authoritarian state, they were at the center of what I have defined above, following Peter Wagner, as the "discursive coalition."[6]

The imperial discursive coalition

This arrangement between the political and academic elites was not static, but rather an alliance that continually underwent modification as the social field altered. Any of the discourses and institutions of the main participants, scientists and politicians, could initiate change, as could other elements of the field. Social scientists simplified the complexities of the time by formulating well-demarcated "entities" to establish the boundaries of the coalition and to resolve what they perceived as crises arising within the larger field. In so doing they presented themselves as the voice of "science" and advocated solutions that supported the projects of certain political actors. The institutions created by those actors in turn served to legitimize and underwrite the activities of the social scientists, hence the coalition. Rules for scientific procedure were aligned with philosophies about the education of the cultivated public (*Bildungsbürgertum*) and about the relationship of the state to other institutions, including economic ones. The scientist, as a university professor, was also an official of the state and a leading figure in the cultural circles where public opinion was shaped. The integration of these roles gave him symbolic power as the interpreter of the cultural code of mainstream German society, the bridge between the political establishment and the larger public. This power was the source of his cultural capital, which he took care not to squander. It was his "fiscal" conservatism that earned him the label of "mandarin."[7]

German social scientists tried to integrate several different roles.[8] As academic specialists, most considered themselves to be objective scholars and many, like Alfred Weber early in his career, were strongly committed to empirical research. This specialization meant narrowing the scope of their scholarly endeavors, which weakened their effectiveness in three other roles. While some were content with this restricted role, others felt great tension between their narrow research agendas and their desire to be public intellectuals defined by those other roles. The second role was that of "cultural citizens" (*Kulturbürger*), members of wider cultivated (*gebildet*) public. As such, they were active and important participants in formation of public opinion to the point of articulating cultural values for the larger community. Thirdly, they were participants in the political process with varying degrees of commitment to different political forces. Many saw themselves as "unpolitical," "politics" being defined very specifically as the pursuit of domestic partisan, and especially political party, interests. These "unpolitical" men were in fact political in the broader sense in their strong support of the authoritarian bureaucratic state and accompanying nationalistic rhetoric. But some, like the Webers, became involved, if only briefly, in the parliamentary process by actively affiliating themselves with political parties. Finally, they were diagnosticians of their time, meaning that they sought to provide principles and values by which their age could orient itself. Although individual social scientists placed different emphasis on each of the four roles, I agree with Volker Kruse that many, including Alfred Weber after the turn of the century, saw the fourth as the most important. Those who took this task as diagnosticians seriously saw the need to address the issues that arose within the reform milieu. However, this role cannot be examined in isolation of the others, because part of delivering a successful diagnosis involved understanding the relationship of the other three roles to one another.

At the turn of the century, establishment social scientists tried to develop language and concepts to reconcile oppositional forces with the institutions of the discursive coalition. This strategy advocated varying degrees of modification of those institutions, and it was not without its own conflicts about the extent of those changes. While most of this discussion took place within the accepted universe of discourse without any challenge to the reigning presuppositions, a minority began a structural alteration that would call into question the ruling set of binary relationships. As we shall see, the lines drawn in this challenge were largely generational. As an important member of the oppositional generation, Alfred Weber's agency eventually helped to redefine the priorities of the system, but he never completely discarded the old habitus. Its structural assumptions continued to shape his thought even after the imperial coalition itself had disappeared. The next section examines that structure.

The binary construction of the discursive coalition

Throughout most of this period, the discursive coalition was organized according to binary constructions, as were most competing discourses. The more intense the pressures for change, the more evident the binary structures became as a means to incorporate that change into discursive strategies of orientation. Among the dueling dualisms that organized the various discursive formulas were: *Gemeinschaft* (community) and *Gesellschaft* (society), state and civil society, culture and civilization, education as cultivation (*Bildung*) and education as training (*Erziehung*), spirit (*Geist*) and matter, spirit and soul (*Seele*), organism and mechanism, human sciences and natural sciences, historicism and positivism, synthesis and specialization, bureaucracy and parliament, notables and parties, and, most crudely, heroes and traders. German social thought (the discourse in both its noninstitutional and institutional forms) made use of many of these dualisms with varying degrees of sophistication out of the need for orientation in the rapidly developing industrial society. The more a crisis was perceived (as during World War I), the more evident the binary constructions. Thus, the dualisms were more implicit than explicit in the first of the roles delineated by Kruse—that of specialist—especially if the author did not seek to provide some kind of larger orientation. Conversely, they were strongest in the role of diagnosticians of the times. But, regardless of the degree of its explicitness and the specific content it organized, the binary form prevailed within the German academic habitus and within the larger cultivated public.

Appreciation of how the binary form organized the discursive coalition in the late nineteenth and early twentieth centuries requires more than a list of dualisms and the assertion that the degree to which they could be found in someone's work can be correlated with that person's need to address the problem of orientation. Viewing these binary concepts as part of an unchanging list from which intellectuals could select makes them too stable. For example, Raymond Geuss has noted that two of these dualisms, culture-civilization, and cultivation-education, were not always binary constructions. At the beginning of the nineteenth century, the second term was a "shadow" of the first, originally almost synonymous and overlapping. Only later did it become antonymous.[9] Instead of a stable list, the binary terms should be seen as forming an unstable constellation, to use one of Alfred Weber's terms. Depending on the thinker, the configuration of the elements differed. The various discourses interpenetrated one another and interacted with one another, with the result that concepts and terminology remained fluid. Adopting one binary formulation might demand ignoring another or even denying its validity. Or one might retain a dualism but reassign the elements organized by that dualism. Such practices incorporated the strategy of euphemism discussed above.

This shifting constellation of binary constructions forms an important part of my interpretation. I argue that the confusion or disagreement about the nature of the binary elements within the discursive coalition in which the German human sciences, and more specifically the Historical School of National Economy, participated at the turn of the century helped trigger a feeling of "crisis." The third generation of the Historical School, which included Alfred and Max Weber, in responding to the crisis and by so doing accentuating it, did not simply reject the entire discourse in which they had been schooled. The binary form remained central for them, but the evaluation of the content organized by that form did not. They objected to certain elements of the coalition, which made them more receptive to outside elements and in turn led to a reconfiguration of the elements organized by the primary binary structure. Their new dualistic structure thrust into question the relationship of the university to both the monarchical, bureaucratic state and the public sphere of opinion that supported it.

Politics and publics

The central dualism that organized the discursive coalition through most of the nineteenth century in Germany was that of the state and civil society (*bürgerliche Gesellschaft*), which assumed its most philosophical form in Hegel's work. Within the coalition, this dualism provided legitimization for the authoritarian, bureaucratic state. The role it played is illuminated by contrasting it to an alternate model, the parliamentary state that was dominant in Britain. Jürgen Habermas has depicted the "parliamentary" and the "bureaucratic" forms as political ideal types, accompanied by a corresponding "public sphere."[10]

The public sphere associated with the parliamentary type represented the attempt of civil property owners to protect their property from the encroachment of the authoritarian state of the early modern period. They formed a civil public sphere, which they assumed to be a place of rationality governed by natural laws, embodying the principle of legal order, and shaped through critical debate. The members of this public sphere saw themselves as the arbiters of legal and moral issues. In formulating a public opinion, they conflated their roles as human beings who communicated through critical debate in the world of letters and as property owners who communicated through rational-critical debate in the political realm institutionalized in parliamentary parties of notables. Thus, the material interest position of "bourgeois" was identical in their minds, if not in reality, with the political position of "citizen" and the moral position of "person." Two different groups of actors, the economic/political and the intellectual, presented themselves as a single entity, the public. Men of letters, who as vendors in the marketplace of ideas articulated public opinion, considered themselves part of the same system of public communication as the political representatives in parliament.[11]

In contrast to the parliamentary type, the bureaucratic type rejected the idea that political values emerged from critical debate in the civil public sphere. Proponents of the bureaucratic ideal characterized this public sphere as divisive civil society, where reason no longer realized itself, and, therefore, as necessarily subordinate to the authoritarian state, which offered a synthesis of spirit.[12] Part of the state, the bureaucracy, assumed the rational faculty that had been assigned to the public sphere in the parliamentary type. Officials saw themselves as the upholders of the state of laws (*Rechtsstaat*), which provided rational restraints on the arbitrary nature of both the whims of the monarch and the forces of civil society. Cultivation (*Bildung*), the course of education that qualified officials for their posts, also made them leaders of a larger cultivated citizenry (*Bildungsbürgertum*).[13] This type also assumed the leadership of notables, but unlike those who participated in the mechanistic conflict of private interests that characterized civil society and were represented in the parliament, these notables were seen to represent a common organic unity of spirit and stood above parliament.

In summary, the parliamentary version assumed that the public sphere consisted of a unity of property and letters and their respective elites (notables), was *separated from monarchical authority*, and thus was represented politically through the institution of parliament. The bureaucratic version drew the dividing line in a different place. The sphere of letters (cultivation) was *joined with monarchical authority* through the mediating stratum of the bureaucracy and, hence, was separated from the material interests of property owners and the institution of parliament. Not only was the creation of meaning shifted from the civil public sphere to the authoritarian state, but the relationship of that dualism was conceived as *hierarchical rather than oppositional*. During the course of the nineteenth century, the bureaucratic type proved to be the stronger as Germany made the transition from a politically divided "cultural nation" to a unified "national state."[14] Although the parliamentary type did not disappear, it suffered a series of defeats during the century and survived only in its "weak" form. Nowhere was this hegemony of the bureaucratic type more evident than in the German universities.

That model placed great importance on the German professoriate, who educated members of the bureaucracy and were, at the same time, officials themselves. As members of both the state and the cultivated public sphere, professors formed a link between both, a position that bestowed considerable cultural capital upon them. The scientific discourse of the professoriate was combined with the ideal of cultivation and the political discourse of the German state to form the bureaucratic discursive coalition.[15] As officials, professors conducted their intellectual projects with an inseparable mixture of interest in knowledge and service to the state. Accordingly, they idealized the state as the embodiment of the spirit of the nation (*Volksgeist*), standing above the particularity of social interests, and thus as a "higher ethical power that dominates individual existence." Academics such as the historians Sybel and Dahlmann believed that the state represented "the actualization of freedom through the power of the community (*Gemeinschaft*)" and that "one cannot be a nation (*Volk*) without a state."[16]

Historians, who recounted the deeds of great statesmen as the embodiment of the moral development of the nation, were the leaders of public opinion during the middle of the nineteenth century. However, with the tremendous economic changes of the last third of the century and the emergence of significant oppositional forces, leadership in the formulation of the discursive coalition passed from the historians to the national economists.[17]

The Historical School of National Economy

This change in leadership did not mean an abandonment of a historical perspective, as the Historical School of National Economy dominated that discipline within the German university. The Historical School, which first appeared as a counter force to the classical political economy of Adam Smith and David Ricardo, reached its height during its second generation under Gustav Schmoller, who was the preeminent figure in German national economics from the founding of the empire in 1871 until World War I.[18] Schmoller's power was especially strong in the 1890s, when to his editorship (since 1881) of the most important economic journal, the *Journal for Legislation, Administration and National Economy in the German Empire* (known as *Schmollers Jahrbuch*), he added in 1890 the chairmanship of the practical wing of the school, the Social Policy Association (*Verein für Sozialpolitik*), and in 1897, the rectorship of the University of Berlin. During this decade, Alfred Weber began his academic career under Schmoller.

The discursive system of the Historical School (and Schmoller) contained five main characteristics.[19] First, it sought to replace the abstract theories of classical economics with the detailed description of economic facts that resulted from a direct observation of economic reality as it changed historically. Schmoller rejected the notion that a complete and closed economic system could be constructed out of an abstract theory of human nature disconnected from reality.[20] He attributed this incomplete position not only to classical economics but also to Marxism and the Austrian School of Marginal Utility.[21] In place of these abstract speculations, Schmoller advocated "more rigorous" methods, especially statistical surveys like those carried out by the Social Policy Association. After years of detailed and specialized research, "national economy has ceased to be a liberal art. It has become, like others, a specialized discipline."[22]

Second, for the Historical School, the economic process was not a separate and fully automatic mechanism as it was for their rivals, but rather was embedded in an organic "flow of historical becoming."[23] This explains the seeming contradiction in Schmoller's damnation of classical economics as a "closed system" while praising German national economy for trying to get a picture of the whole as a "closed unity." The latter did not seek universal laws, but remained tied to the historical destiny of the German nation.[24]

> A sum of unified feelings gives the nation (*Volk*) its soul; a sum of unified representations crosses over the threshold of the national consciousness and produces that which we call the unified spirit of the nation (*Volksgeist*). The latter expresses itself in unified customs, strivings and acts of will, dominates the doings and drives of all individuals, including those of an economic nature.[25]

Importantly, in delineating this organic unity, Schmoller posited no antithetical relationship between the terms "spirit," "soul," and "will," the latter two both being seen as complementary to the more central spirit.

Third, the state was viewed as the greatest representative of this organic spirit of the nation. In Hegelian fashion this position of the state was juxtaposed to the divisive interests of classes and the parties that represented them. Accordingly, Schmoller praised the monarchy and bureaucracy as neutral elements standing above social conflict, and he deprecated parliamentary democracy as promoting disorder.[26] At the first meeting of the Social Policy Association, in 1872, he summarized the views of those gathered thus:

> As proof of their sincerity for the constitutional system [of the German Empire], they do not want the alternating class rule of the various economic classes, which combat one another; rather they want a strong state power that stands above the egoistic class interests, that legislates, conducts its administration, protects the weak, and elevates the lower classes with a just hand.[27]

This "just" objectivity also applied to academics, who were officials of the state. The constitution of the university allowed professors to be concerned only with the general welfare by making them independent from economic class interests, for which there was no place in the university.[28] Schmoller's position was the definitive statement of the bureaucratic discursive coalition.

Fourth, national economy had a strong moral element, where "morality" (*Sittlichkeit*)[29] meant establishing a positive relationship between the individual and the whole. When the state, the community and the personality worked together, the result was a "healthy moral national spirit," which served as the basis for national economy. Thus economics could never simply examine the instrumental activity of egoistic individuals, but always had to be tied to the ethical life of the nation, especially as it was expressed through law.[30]

Fifth, this interpretation of the state as the embodiment of the national spirit and morality was reinforced by the perceived failure of classical political economy to come to terms with the economic and social realities of industrialization. The more concrete historical approach of the Historical School demonstrated the necessity of state intervention to redress these shortcomings. The "social question," the issue of how to deal with the growing masses of industrial workers who were increasingly becoming politically organized, was to be answered with a program of reform that would incorporate the working class into the national unity.[31] National economists, said Schmoller, had the duty "to raise, to cultivate and to reconcile the lower classes so that they will become a part of the organism of society and the state in harmony and peace."[32] Therefore, a philosophy based on the moral activity of the state, on public care through legally regulated principles, was superior to one based on rational self-interest. Members of the Historical School saw their scientific efforts working in two directions: they contributed to the cultivation of those officials who were actually carrying out a program of economic and social reform, and they helped to shape a wider cultivated public opinion.[33] The result was "Social Policy" (*Sozialpolitik*), which represented the transcendence of the divisions between society and politics, the synthesis of the two discourses of the bureaucratic state and historical social science.[34]

In summary, the Historical School took an inclusive approach to the fields impacting and impacted by the discursive coalition. It connected (and subordinated) economics to the larger social, moral, and governmental picture. Accordingly, the school's practical arm, the Social Policy Association, was viewed as a scholarly forum, a shaper of public opinion and a resource for the bureaucratic state.[35] This set of organic presuppositions about the synthetic centrality of the state informed its members as they proceeded with their detailed empirical studies.[36]

The Social Policy Association's articulation of the discursive coalition was expressed in binary terms, with an emphasis on the hierarchical dualism of state and civil society. A series of other dualisms complemented this pair: the state represented the organic realm of spirit where primacy was placed on the moral duty of the state official; civil society was the mechanistic material world, where emphasis lay on individual interests promoted in parliament. Such a division meant that the human sciences had to remain historical in their methodology and not adopt the abstract positivist methods of the natural sciences, the latter being best restricted to physical matter. Specialization and statistical studies were beneficial, provided they remained subordinated to the larger synthetic morality, just as civil society had to be subordinated to the state. In this sense these academics were good bureaucrats, working out the details and accepting the overall orientation provided by the leaders of the state. As Schmoller put it:

> All great ideal possessions of humanity, Christianity, the development of the law through the millennia, the moral duties of state power, above all as they have developed in Germany and Prussia,

point us to the same path of reform that the imperial messages of 1881 and 1890 prescribed for us. German science has done nothing other than attempt to provide a causal grounding for these ancient ethical-religious and legal-governmental imperatives and to produce rigorous evidence of their truth.[37]

The challenge to the Historical School

During the two decades before World War I, this discursive coalition was challenged on a number of fronts. Volker Kruse, Gunter Scholtz, and Helmuth Schuster describe the three most important challenges. Kruse emphasizes the methodological opposition to the Historical School by highly theoretical approaches, namely positivism, Marxism, and the Austrian School of Marginal Utility. He argues that the third generation (especially Tönnies, Sombart, and the Weber brothers) felt the need to reconcile the abstract, nomothetic approach with the empirical, ideographic one.[38]

Kruse is correct, but he does not explain why Weber's generation felt this need for reconciliation more strongly than did Schmoller's. After all, these challengers were not really new; in fact the historicist world view had arisen to contest their predecessors. All the challengers were seen as continuations of extreme Enlightenment thought and, thus, fit neatly into the dualism of abstract theory versus empirical reality upon which the Historical School was founded. Marxism, especially, was dismissed as a party doctrine and, therefore, lacking the objectivity that characterized academics, who as officials shared the general view of the state. As Schmoller explained:

> One dams up the progress and development [of science] if one presents decaying and obsolete tendencies and methods as equal to superior and better developed ones. Neither the strict Smithians nor the strict Marxians can today make the claim to be held as valid. He who is not grounded in current research, current academic cultivation, and methods is no useful teacher. The same is true for representatives of economic class interests.[39]

The challenge of abstract theory in itself was not enough to instigate a feeling of inadequacy among the adherents of the Historical School because it did not disrupt the latter's binary formulation. As a result, the challengers were unable to institutionalize themselves in more than a marginal way within the German university "system." They remained confined to the systemic "environment," and their presence contributed to the definition of the system's boundaries. Gunter Scholtz argues that the most fundamental challenge to the Historical School came not from external challenges but from within, namely in the transformation of the human sciences from a philosophical approach to one of historical empiricism, in which empirical research became increasingly disconnected from metaphysical presuppositions and normative pretensions. Historical national economy was an exemplary discipline in this process. Its emphasis on empirical research contradicted the traditional role of the human sciences—providing orientation for the larger public. As disciplines became more specialized and more empirical, as language became more technical and differentiated from that used in public cultural discourse, a gap between professors and other cultivated citizens grew, making it difficult for academics to provide the kind of synthesis of values that they felt they should be offering to society at large.[40] That orientation could only be supplied by a philosophical, or theoretical, approach. In the same vein, Rüdiger vom Bruch has written that the great reputation of German professors in the nineteenth century came from two different developmental lines: the "scientization" of disciplines leading to specialization and their role in the bourgeois public sphere as the spokesmen for the moral-cultural community.[41] But the two lines could be contradictory. The specialization that increased the international prestige of German scientists also weakened their claim to provide the synthesis that would make them the arbiters of the German nation. Here, one can see growing disparity between the above-noted roles of academic specialist and cultural citizen.

But, again, there was no reason why this discrepancy should cause a feeling of crisis. Specialized scientific discourse did not have to articulate the larger synthesis of meaning. It simply had to be symmetrical with that synthesis. Schmoller argued for specialized studies that he felt were complementary to the larger moral vision, especially when they were attached to Social Policy, the social reform program of the state. By the Wilhelmian period, the resolution of the tension between the roles of specialized science and orientating cultivation had become inextricably bound up with the state, which served as the mediator between the increasingly specialized professoriate and its public. Professors might lament a growing distance from the larger public, but their specialization did not contradict the underlying principles of the discursive coalition.

That coalition was more severely challenged in the 1890s, the very decade of Schmoller's preeminence and Alfred Weber's entrance into it, by major changes in the German state and society. Bismarck left the scene, replaced by men of lesser stature. The *Junker*, the eastern nobility whose lot was identified with the state, demonstrated that they were more interested in their own economic well-being than in that of the nation, by vigorously opposing the attempt to lower protective tariffs as the economy improved.[42] The Social Policy Association studies by Max Weber during this decade revealed the *Junker* to be agrarian capitalists who brought in migratory Polish labor rather than support the more traditional German cottagers.[43] In conjunction with these agrarian changes, industrialization and the accompanying urbanization drew migrants from the east and swelled the ranks of the working classes. The legalization of the Social Democratic Party in 1890, following on the heels of large strikes in the Ruhr mining district, accelerated the party's climb to the electoral plurality, reached in 1912. The SPD's increased importance paralleled the greater importance of the *Reichstag* as the primary barometer of public opinion. In addition, an antimodernist populism grew among small farmers and artisans, and the bureaucratic state found itself facing the "politics of demagogy."[44]

Helmuth Schuster takes note of these types of forces and adopts Habermas's argument that these changes created a new fragmented public. Both the parliamentary and the bureaucratic types, with their restrictive public spheres led by notables, were faced with a new expanded public organized along class lines, which also demonstrated a mutual penetration of the spheres of society and state. New political forces, characterized by the propaganda and lobbying tactics of mass political organizations, avoided dealing with the bureaucracy except at its uppermost levels. These groups made connections between their own public and the state and, in so doing, discarded the older notion of an organic public on which the discursive coalition of state and academic science had been based.[45] Interestingly, the liberal parties, which were supported by the Webers, had the most difficult time adjusting to the deterioration of notable politics.

In response to this new public, Schuster argues, academics like those in the Historical School engaged in two types of studies, "situational studies" (*Lage-Studien*) and "ideational studies" (*Ideen-Studien*), the former being a transitional form that gave way to the latter. Situational studies examined the standing and development of social groups (most importantly the industrial proletariat), whose material deprivation prevented their being integrated into the nation. These studies were empirical and statistical—such as the Weber brothers' investigations in the 1890s under the auspices of the Social Policy Association—treating society as something separate from the public and its ideals, so that it could be "objectively" studied without putting those ideas in question. This approach enabled researchers to single out groups to be studied without viewing them as forming separate publics with their own ideals. Situational studies were conducted in accordance with the bureaucratic discursive coalition and the ideal of Social Policy while studying groups that did not accept those premises. Situational studies, therefore, presented the problems of the new industrial society to the old preindustrial public without really facing the inherent contradictions in that endeavor.[46]

Ideational studies, which came largely from Weber's generation of the Historical School, represented the recognition of the fragmentation of the public sphere into many publics, each with its own ideals. Accordingly, practitioners of ideational studies did not separate the sphere of social groups from that of public ideals. They recognized the inner differentiation of the new public and believed that society constituted itself through class energies and the ideas that bore them. One could not remain above these forces, as the discursive coalition claimed to do, but had to participate in them. Such a position placed the Social Policy of the state (the cornerstone of the discursive coalition for national economists) in question. Ideational studies were part of a tendency of the third generation to look beyond the state and officialdom for a new cultural solution to the disorientation that characterized these years. It would be going too far to say that the national state yielded to a return of the cultural nation, but clearly the latter played a much more important role than it had. "Culture" had once again become an important term for thinkers like Alfred Weber as the sphere where the fragmentation of modern society could be overcome, as the location for a new meaningful unity.[47]

In looking to culture as the sphere of redemption, the formulators of ideational studies were attracted, in varying degrees, to the "aesthetic" counterculture outside the university. This counterculture ranged from an elitist cultural pessimism to a fascination with the iconoclastic Nietzsche to the resurgence of vitalism (*Lebensphilosophie*) in its various manifestations.[48] None of these movements seriously challenged the organic-mechanistic structure of the discursive coalition, but they disagreed about where elements should be placed in that binary.

Weber's generation of the Historical School viewed the state as a cultural institution, but not as *the* cultural institution that defined the nation. Increasingly, they viewed the capitalist economy as the defining force in the modern world and they sought to examine capitalism not simply as an economic phenomenon but also as a cultural one.[49] The two most important statements of that position, both of which can be seen as ideational studies, were Werner Sombart's *Modern Capitalism* (1902) and Max Weber's *Protestant Ethic and the Spirit of Capitalism* (1904). These two built on Ferdinand Tönnies's seminal 1887 work, *Gemeinschaft und Gesellschaft (Community and Society)*. In these studies the central binary hierarchy of state and civil society, the underlying premise of the bureaucratic discursive coalition and Social Policy, was replaced by the dualism of Tönnies's book.

All of the most important members of the third generation of the Historical School made this binary switch, which retained the organic-mechanistic dualism of its predecessor, but refused to conflate it with the state–civil society dualism.[50] Instead, they moved the German state (including its officials) from the organic side of the dualism to the mechanistic side. No longer was the state the "objective" embodiment of the national spirit, but it was rather a part of the increasingly rationalized "*Gesellschaft.*" After Tönnies included the nation-state with capitalism among the forms of *Gesellschaft* in 1887, most of these thinkers made this conversion, which allowed them to make capitalism the defining element of their era. No writing was more typical of this strategy than Alfred Weber's study of state officials.[51]

To take such a strategy and substitute *Gesellschaft* for civil society meant viewing the world in a structurally different way. Instead of a *hierarchical* model in which an organically unified realm (the bureaucratic state), stood above as a synthetic unity and provided order and meaning for a divisive, mechanistic one, the new model was *successional*. The organic traditional realm did not stand above the mechanistic modern one, but gave way to it in all its forms.

In summary, for most of the nineteenth century the bureaucratic discursive coalition successfully combined three elements: the authoritarian, bureaucratic state; the cultivated public; and university social science. Professors in their three roles as state officials, cultural arbiters, and scientific specialists saw less need to fulfill the fourth role, that of diagnosticians of the age. The success of the coalition was based on the ability of all of these elements to reinforce one another. As the ties between the political and cultural notables and the larger public began to disintegrate, academics proved to be increasingly ineffective in their mediation between the two. This made some more receptive to the socioeconomic theories and the cultural criticism that previously had challenged the coalition unsuccessfully. In the

1890s, Schmoller's generation maintained the façade of the discursive coalition, but the generation of the Webers, with long-standing ties to parliamentary forces and less stake in the existing formulation, began to look away from state activity and Social Policy toward broader, more cultural ways to address the newly fragmented public.[52]

The challenge of the third generation burst forth in the decade prior to World War I. There were open and heated confrontations in the meetings of the Social Policy Association, with the Webers and Sombart leading the way. To counter Schmoller's influence, they established a competing journal to his *Jahrbuch*, the *Archive for Social Science und Social Policy*, edited by Max Weber, Sombart, and Edgar Jaffé. Less successfully, they countered the Social Policy Association with a new institution, the German Sociological Society, which was intended to divorce scientific theory from the practical orientation of the Association. And, with still less success, they supported Friedrich Naumann's efforts to create a viable parliamentary party.[53]

This challenge did not represent a total break with the preceding generation. The Webers' generation retained elements from the established discourse, including its binary structure. But they were also willing to consider oppositional elements that had been previously rejected out of hand. Most importantly, they abandoned hierarchical dualisms in favor of new successional ones, disrupting the whole structure of the old discourse. Those who adopted the new structural strategy had three options for facing this new world. They could adopt the cultural pessimism of decline that was in vogue outside the university.[54] Or they could seek a new type of meaning in the terms of the modern rationalized world, as did Max Weber. Or they could hope to restore an organic unity to the world without trying to turn back the clock in a reactionary manner. It was this third approach that was adopted by many in the third generation, including Tönnies and Alfred Weber.[55]

As indicated above, Weber's generation of the Historical School did not simply begin their careers by writing ideational studies. Those came only in the first decades of the twentieth century. In the 1890s, they cut their teeth on the situational studies that typified Historical School activities of that decade.[56] In conducting these empirical studies, they became aware of the fissures in the discursive coalition. Of the important figures of his generation in the Historical School, Alfred Weber, the youngest, was the last to fully challenge Schmoller. His early studies for the Social Policy Association were well within Schmoller's premises. Yet, even here, one can find indications of the changes that were to come.

2
Early Economic Writings, 1897–1908

At the beginning of Alfred Weber's academic career, from 1895 (when he finished his doctoral dissertation) until 1904 (when he accepted a position as full professor in Prague), his work was situated firmly within the discursive coalition. From 1900 until his call to Prague, he taught as an untenured lecturer under Schmoller at the University of Berlin. Although he had already begun to show some signs of independence during this period, this early work was conducted under the direction of Schmoller within the framework of the Social Policy Association.[1] In 1897, Weber's doctoral dissertation was published as an article in Schmoller's *Jahrbuch* as was his *Habilitationsschrift* in 1901. These clearly fit Schuster's description of "situational studies" as transitional works. While one can see some discrepancies with the views of the Historical School of National Economy and the larger discursive coalition, they operate within its framework in that they look to the state as the main agent in transcending the problems of civil society, especially the "social question."

The first clear break with Schmoller's generation of the Historical School came in the first decade of the twentieth century as a result of Weber's stay in Prague. There he produced an abstract economic treatise along the lines of the rival Austrian School of Marginal Utility. However, this diversion would be only temporary. By the end of that decade, back in Germany, he began to write ideational studies of the kind that made his generation so important.

Empirical studies and the Social Policy Association

The most important issue for the Social Policy of the Historical School was the "social question," the role of the working classes in the German community. This was not simply an economic question but also a cultural/moral one. How did one overcome the economic forces that prevented the working classes from being integrated into the national unity? Weber's early studies addressed this concern. One theme dominated all others in his writings during this early period, the role of domestic, or putting out, industry (*Hausindustrie*) in the German economy, especially as evidenced in the production of ready-made clothing (*Konfektion*).

The relationship between factory labor and other forms of labor such as craft artisanry (*Handwerk*) and domestic labor, had become a very important issue in the mid-1890s. In 1894, Gerhard Hauptmann's drama *The Weavers*, viewed as a protest against the conditions of domestic laborers, was produced for the first time. Two years later, a large strike occurred in the textile industry, which was largely organized around home production.[2] As a result, the Social Policy Association decided to investigate domestic industry, and a survey was distributed. Weber was a leading member of the commission that organized the study.[3]

Weber's work on domestic industry drew him into issues concerning the binary relationship of traditional society to modern industrial society, a primary concern of the discursive coalition. In labor relations this concern was often expressed as the relationship between the artisanal crafts and large-scale industry. In the simplest formulation, artisanry was characterized by an organic unity of ownership and labor. As a master, the artisan performed both these roles; and even journeymen and apprentices were viewed as part of the master's family and, in effect, as developing themselves to be masters. Such a view set up a dualism in which this organic artisanry was depicted as the binary opposite of mechanistic factory labor. Most of those who investigated domestic industry, and tried to fit it into this binary schema, viewed it as a transitional stage between the two other forms. But which set of values predominated? Did domestic workers, because they worked in their homes or in small workshops, share the same values with the traditional artisans of the craft system? Or were they in fact a primitive version of the modern proletariat? As Weber noted, most investigators leaned toward the former view:

> [Two] fixed general ideals antagonistically confront one another in the observations and judgments [about domestic industry]. The one sees domestic industrial workers even today as the ideal workers in large industry, because this man, although inside modern industry, still represents the beneficial type of the old artisan that

is external to it, and in him one can presuppose the elements of the artisan's character. The other perspective, more from the standpoint of the working class, sees in the domestic worker a retrograde element, which by his acceptance of existing conditions is a millstone around the working class's neck and restricts its progress.[4]

Here, Weber phrased the issue in the conventional binary terms. Both traditionalists and the modern labor movement saw domestic industry as a residual form of labor that preserved traditional values, and they judged it accordingly. In his writings of this period, Weber, while generally siding with modern labor, would overturn the conventional wisdom of both sides and describe domestic industry as the product of urbanization and the resulting oversupply of cheap female labor and not as a residual element from an earlier world.[5] Thus, domestic industry was placed on the modern side of the traditional-modern dualism, and was presented as actually destructive of traditional values. Even if Weber did not realize it at the time, this was an important precedent for the binary shift that would occur a decade later, when he would move state bureaucracy to the mechanistic side of the dualism.

Weber described three types of domestic industry. In "pure domestic industry," the worker was independent of outside forces in both sales and production. This type of labor, which involved no middlemen, was simply an extension of traditional household activity in response to the growth of (usually local) markets. The worker had all the benefits and uncertainties of an independent producer. Weber did not explain how this group differed from artisanry, but one could assume it was due to a lower level of skill plus the absence of any organization such as guilds. Weber believed that this "pure" type was largely benign and might even be constructively preserved in the form of cooperatives. He did not direct much attention to this form, because it largely belonged to an earlier, preindustrial stage of production and, thus, increasingly yielded to the other two types of domestic industry.[6]

Those two types Weber termed "external labor," which was directed in both production and sales by external entrepreneurs, and "retail labor," which was completely independent in the production process but dependent on an outside sales organization. He believed that retail labor was a transitional stage between pure and external domestic industry. Some of these producers might be raised to the position of independent small producers, in other words reverting to the pure type or becoming artisans, but most would become modern industrial workers, in the form of either external domestic workers or factory workers.[7]

External domestic labor most concerned Weber, for it was the predominant type in the industrial capitalist era. It was entirely modern, the product of a later stage of domestic industry, with no traditional component. When Weber railed against domestic industry, it was this type that he had in mind. External domestic labor differed from industrial factory labor only in the location where the work was performed. These workers had no independence; they were completely controlled by their employer. They did not produce for local customers as the pure domestic worker and artisan did. Rather, they were simply cogs in a large organization that sold goods in anonymous external markets. They, too, were merely appendages of their machines. Weber wrote:

> The old domestic industrial worker was the lord of the labor process that he undertook in his small house or workshop. He oversaw this labor process and enabled it to independently develop further, for he or people like him had raised it to the level at which it stood.
>
> The new domestic industrial worker, left alone in his individual workshop by modern technology and the decomposition of labor, no longer oversees the labor process. A position above him has organized and reconfigured it. It has also reconfigured the product that is manufactured. Both alter themselves without interruption and without the worker being asked. This new domestic industrial worker is hidden in an invisible machinery that moves him according to its will, in which he, like his colleagues in the factory, is only a cog. He conducts himself in relation to this machinery and to the work that it allots him exactly the way the factory worker does, that is, passively. He is no longer the lord, but rather the servant of his labor.[8]

Weber claimed that although domestic labor was a form of modern industrial labor, it was much more disadvantageous to the worker than the other modern form, factory labor. The latter benefited from the protective legislation, including policing, that had been passed since 1890. Despite the fact that this legislation applied to all workers, it was much more difficult to monitor industrial practices when these were dispersed into individual homes or small workshops. In addition, the income of factory workers was higher than that of domestic workers.⁹ It should be noted that Weber hardly mentioned another advantage of factory labor over domestic labor—its ability to organize. This emphasis on protective legislation over collective bargaining reflected the attitudes of most of Schmoller's generation of the Social Policy Association. It privileged actions of the state over negotiations within civil society.

Weber believed that modern market forces were responsible for the prevalence of domestic industry. He detected a clear relationship between the supply of labor and the form of production. When demand for labor exceeded supply, the productive process was more concentrated, the epitome of this form being the factory. When supply exceeded demand, then production forms were dispersed, the factory manager being replaced by the middleman in the domestic system. Domestic industry, therefore, depended on an oversupply of labor, mostly unskilled, often part-time and, in industries such as ready-made clothing, usually female. Weber believed that migration of population from the countryside into the cities created this oversupply. With the rising cost of living, the high cost of land where factories and workshops could be built, and the inability of the demand for household help to absorb the new arrivals, the conditions were present for large city domestic industry. In addition, the machinery that was introduced into these industries, especially the sewing machine, could be installed in the home, not necessitating factories. This combination of production conditions attracted domestic industry out of the countryside and actually expanded it. External labor also remained prominent in the mountainous countryside, where lack of transportation kept out competition from other industrial forms and maintained the oversupply of cheap labor.[10]

The only solution to the problem of the modern form of domestic industry, Weber argued, was to work for its elimination. This was not to be done all at once, for that would create dislocation among domestic workers. And as indicated above, Weber held that although some domestic industry could survive,[11] external domestic labor had to be targeted for elimination. His commission proposed a number of legislative measures to accomplish this. Improved transportation in both cities and the countryside was important, for it provided access to areas on the edge of cities where the lower price of land would reduce the cost of building factories. In the countryside, where land was cheap, an increase in rail lines would allow better connections with markets, which would encourage the building of rural factories. In addition to reducing the cost of factory construction, the pool of labor had to be reduced, which would increase wage levels and make domestic industry less attractive to entrepreneurs. Weber believed that the growth of white-collar office work would supplement household service as a competitor for the labor of women. He also advocated a strong legislative program, consisting of three components, to eliminate this oversupply: first, the creation of a tax on domestic piecework, which would raise the price of labor; second, the extension of welfare benefits to include those unemployed by the curtailment of domestic labor, which would also remove people from the labor market; and third, the denial of access to certain types of domestic labor to women and children. The committee strongly advocated that those married women who were bound to the home be excluded from the part-time domestic labor market, for they could not easily switch to factory work.[12]

Weber presented the Social Policy Association's approach to reform as superior to those of both traditional middle class groups and organized (socialist) labor. The socialist movement, he wrote, saw protective legislation as "theoretically the most effective way to strengthen the working class, which in its view was threatened with bodily and spiritual degeneration." It did not distinguish between different types of labor, wanting to free them all from the oppression of the higher classes. The middle-class anticapitalist movement was motivated by its deep dislike of modern industry and commerce, which, it believed, weakened family life and religious traditions. Accordingly, it focused on banning work on Sundays and regulating women's labor.[13] Both of these approaches contained a cultural, and therefore moral, argument in addition to the economic one.

In making the case for the Association's position, Weber also addressed cultural issues. He insisted that capitalism was compatible with the cultural development of the nation as long as its crasser abuses were eliminated. He stated that these reforms would be good not only for labor but also for business and the nation as a whole, and he urged German industrialists to drop any opposition they had to this protective legislation. Above all, what this movement "demanded from legislation was simply 'the carrying out of moral principles in public affairs,' as already formulated by Schmoller in 1864."[14] This portrayal placed the Social Policy Association in a superior position to the other reform movements, being neither class oriented nor reactionary, but rather looking toward the good of the nation as a whole. The campaign against the domestic system was presented as a defense of morality rather than an attack on it. At a meeting of the Social Policy Association, Weber argued:

Domestic industry continues to exist here, because this cheap labor power is still present, because it can lower the price of its product through the cheapness of labor. The lower price balances off the lowering of productivity by this method of production and, in the eyes of the public, makes up for the inherent lack of quality of the goods. Is its continued existence therefore justified? That would perhaps be the case in a state whose economic task with regard to the quality of its people can consist of nothing but the production of rubbish with the help of low wages. However, a state that is able to make something else out of its people, because of the cultural level of its population, will in the long run lose out in production using inferior wages to undercutting by countries that are lower down.... There is mercifully no place for a form of labor that rests on inferior wages and by which there can be no progressive development of the masses. This form of labor restricts a part of the nation to a level at which life not only has no value but also offers them no future. If this form of labor is not overcome, it will be a source of danger to the nation.[15]

This stance can also be seen in Weber's other articles of this period that did not deal directly with domestic industry. In an article about unemployment, he argued that economic crises no longer had simple causes, such as a war or a bad harvest, but were periods of readjustment in the economy. This adjustment had become increasingly complex in the modern capitalist economy and involved rectifying an improper division of productive forces. Although Germany had reorganized in one sense with the cartelization of industry, these cartels had not significantly improved the organization of productive forces, that is, of labor.[16] Clearly the situation in domestic industry was an example of such a need for economic readjustment.

As did his brother Max, Weber attacked economic measures that he felt protected the traditional elites at the expense of the nation.[17] He argued against tariffs, claiming that they were different from an earlier autarkism. The earlier policy was designed to make small politically autonomous bodies into ones that were economically self-sufficient as well. Such was not the case with the current tariff legislation, which aimed to protect a single class at the expense of the German nation. Because Germany did not have a great imperial domain, a free world market was more beneficial to the nation than were protective tariffs. The latter were a sign of economic stagnation, not of economic development.[18] For the same reasons, Weber strongly advocated Germany's forming a customs union with other central

European countries, especially Austria.[19] This argument would be developed with a strong cultural emphasis in his later works.

In some ways, Weber's writings of this period clearly reflect his membership in the Historical School. As a member of the Social Policy Association, he subscribed to its program of reform, which included state legislation. Like Schmoller, he assumed that Social Policy had a moral dimension, that the state's role was not simply to monitor economic relationships but also to elevate the moral and cultural level of its people. In these situational studies, Weber indicted domestic industry for being an institution that impeded moral growth.

Despite his agreement with Schmoller on these basic points, Weber demonstrated some divergent tendencies in these early essays. He wrote that the older organic form of labor had given way to the modern capitalistic one, and that one had to acknowledge and adjust to that fact. The problem of domestic industry had little to do with traditional organic forms versus modern mechanistic ones. Rather it was a question of competing capitalist organizations. The old binary formula was discarded as irrelevant to this issue. This represented a step toward challenging the dualistic state-civil society formulation. He took a similar stance when he sided with Max and attacked traditional groups, such as the nobility, who were identified with the state and who portrayed themselves as interested only in the general good of the nation. In actual fact, they represented only their specific class interests. The hierarchically privileged position of groups identified with the state stood at odds with reality. Going further, Weber indicated that capitalism was the revolutionary force of the nineteenth century to which nations had to adjust. Those nations that did not respond to changes in the economic sphere would be left behind in the development of world history.[20]

As we have seen, one of the questions that Weber attempted to answer in his studies of the domestic clothing industry was why ready-made clothing production assumed its organizational form in Germany's large cities. The most important factors determining that location were transportation and labor supply. He would continue to address these very same issues in the coming years, but in a much different manner. He would depart from the empirical methodology of the Historical School's situational studies and adopt that of the School's abstract, positivist challengers.

The location of industries

In 1904, Weber accepted the call to become full professor of national economy at the German University in Prague. He remained there until 1907, when he was called back to Heidelberg as full professor of national economy. In Prague, Weber had contact with members of the important Jewish literary intelligentsia, including Max Brod and Franz Kafka, as well as with Czech intellectuals such as Thomas Masaryk. It was here that Weber became attracted to the philosophy of the French vitalist Henri Bergson, which he claimed helped free him from the ascetic Protestantism of his parental home. In addition to this new appreciation for nonrationality, which would be an important component in his later writings, he adopted briefly the methodology of the extremely rationalistic Austrian school of Marginal Utility. The man he replaced in Prague, Friedrich von Wieser, was a marginal utility theorist who had been called to Vienna, and Weber continued in that vein while there.[21]

The most important product of Weber's stay in Prague was his book *The Theory of the Location of Industries*, his only foray into the realm of abstract economic theory. Because of its high level of abstraction, *Location of Industries* is generally treated in isolation from the rest of his work, which has a strong historical component. This isolation is especially true with regard to his cultural sociology.[22] The book has been the most discussed aspect of Weber's work in the United States;[23] however, its audience has been economists, and not sociologists. In addition to a discussion of Weber's theory, what follows attempts an indication of how some structural elements in the theory foreshadowed certain aspects of the cultural sociology.

There is some direct continuity with the content of the preceding empirical economic writings, if not with their methodology. Weber had dealt with issues of the location in his investigations of the domestic clothing industry, explaining why it had appeared in such strength in large cities in Eastern Germany and why it had remained in the mountainous areas with poor transportation networks. But he had done so within the framework of Schmoller's Historical School, as empirical studies with policy consequences. Now he was moving to the enemy camp. In 1948, Weber offered a brief explanation of why he temporarily moved to such an abstract level:

> The question of the destiny of Germany as an industrial state was a very concrete and, at that time, very burning life-question.... Now there was already a body of theory, which had been extensively worked out, on the "how" of the capitalistic industrial revolution, but not on its "where." And it was on just this point that the question, of whether the strong autonomous tendencies would persist that would allow Germany to accumulate industry or whether they would henceforth be allowed to disappear from the country,

was dependent. However,...in order to gain insight into this "where," one had to first have insight into the purely rational, spatial tendencies of distribution of all industrial production in general, whether it was capitalistic or not.[24]

Weber asked the same basic question that his brother Max did in his agrarian studies: Why did large numbers of the rural populace migrate to the cities? But the answer was very different. Max described the decaying conditions in the rural East, but placed primary causal emphasis on the psychological drive for freedom on the part of the German cottagers who migrated. Alfred ignored psychological aspects altogether, focusing instead on the "general laws," the "pure economic machinery." The intentions of workers were irrelevant to this study. Even the profit motive of the entrepreneur was excluded from the equation.[25] One could argue that Max Weber's situational studies of agricultural workers led to his later ideational studies such as the *Protestant Ethic,* and that both studies address cultural issues but with different kinds of publics in mind. Alfred's study of industry, while it represents a movement away from his earlier situational studies, cannot be called an ideational study, because it is not at all concerned with cultural questions or with addressing a certain public. "We shall exclude from the purview of the pure theory all locational factors of a *purely* social and cultural nature."[26] If one follows Schuster and views situational and ideational studies as stages in the work of the Historical School, then Weber's study of the location of industry represents a step outside of that school toward the rival Austrian School.[27]

Weber acknowledged that his work was a "reshaping" and "developing" of Johann Heinrich von Thünen's pioneering work, *The Isolated State,* written in 1826, which constructed a theory of location for agriculture.[28] The key to Thünen's work lies in the title; in an operation of deductive abstraction, the economic aspects he studied were completely isolated from other factors. Weber's predecessor in Prague, Wieser, explained the procedure as follows:

> Complex experiences cannot possibly be interpreted as wholes. They must be isolated and separated into their elements in order that their effects may be known. The elements, moreover, must ideally be protected from all disturbing influences, in order that the pure effect may be recognized.... Thünen assumes that within [the state's boundaries the conditions of agriculture are uniformly distributed about the central point of a single market—an

arrangement which actually is never met with and can never be expected to be found.[29]

This quotation from Wieser indicates the connection between Thünen and the later Austrian School.[30] And it put his work in opposition to that of the Historical School, whose program was the incorporation of economics into a larger social, moral and political unity.[31] Clearly in this work Weber adopted the methodological side of the Austrian School, if not the details of their theories. He sought out universal economic elements mechanistically organized by generalizing laws.[32] Although his larger plan called for a second volume dedicated to "realistic" studies of particular situations, he never carried out any of those, but simply wrote a general introduction to a collection of studies by others.[33]

The major difference between Weber and Thünen was that the latter, analyzing agriculture, assumed that the location of a farm was fixed and so the question became: What should be produced on that location given its distance from the market? Weber, on the other hand, analyzing industry, assumed that the choice of product was fixed, and the question became: Where should the production "unit" be located given the configuration of locational "factors" and laws that determined the relation of factors to one another? General factors, which were his only concern here, could be reduced to four: the site of consumption, the transportation of materials, the supply of labor, and the agglomeration of units.[34] He stated that in less advanced economies, especially where materials were more or less "ubiquitous," industry was located at the place of consumption.

One could mathematically deduce economic laws that caused industries to diverge from this condition. The most important of these governed the transportation of materials, which applied only when ubiquities were not involved and materials were "localized," that is, located only at certain places. Assuming labor supply and cost to be constant, then a "material index," the ratio of the weight of the localized material to the weight of the processed product, could be mathematically calculated, which would determine the location of the industry.[35] Cost of transportation, determined solely by weight and distance was the most basic factor in the location of industries. If more than one material was used in the process, the situation became more complicated, but the same equation applied. Weber argued that the mechanization of production was generally a "process of weight loss," which strengthened the material components of location and weakened the consumption components.[36]

This basic relationship between consumption and materials became modified when the cost of labor varied from place to place and, in some locations, exceeded the cost of transportation. The relationship of these two factors, given mathematically as the ratio of labor per unit of locational weight, was termed the "labor coefficient." The law of labor orientation stated that a location would be moved from the point of minimum transportation costs to a more favorable labor location if the labor coefficient was high, that is, if "the savings in the cost of labor which this new place makes possible are larger than the additional costs of transportation which it involves." Curves, which Weber termed "isodapanes," could be constructed connecting the points of deviation from the minimum transportation point. Within the "critical isodapane," the economies of labor were greater than the deviation costs. The labor coefficient was affected by factors such as the density of population, the state of transportation, and the mechanization of industry. A greater concentration of labor made it a stronger factor, as did cost-lowering improvements in transportation. Mechanization of a process, which lessened the amount of labor required, weakened the labor factor.[37]

A second conditioning factor, in addition to labor, was the agglomeration of units. Agglomerations occurred at the intersection of critical isodapanes of labor. "Whatever the situation and whatever the quantity of output of any individual unit, if its critical isodapanes intersect with those of enough other individual units to make a unit of agglomeration, it will be concentrated with these others."[38] If there are several of these intersections, the ones with the lowest transportation costs would be chosen. Weber noted that there could be social causes for agglomeration, but he was not concerned with these. Only the interaction with the other general factors could be considered.[39]

Clearly the model for Weber's theory came from the natural sciences, with their mechanistic system of forces acting on atomistic units. Scholars have noted how much Weber's models resemble those of physics; one thinks especially of the Newtonian laws of gravity. At one point, Weber refers to the units as "little balls," and, at another, as "particles."[40] Weber did not view the system as static; he claimed that it was continually modifying itself, much like a physical field would. Moving the location of a unit would, in turn, change patterns of consumption, which would affect the whole system. In fact, he described it as a kind of equilibrium that modified itself when it became partly irrational.

At the same time that he explicated this abstract analytical system, Weber also indicated its limitations. He distinguished the abstract forces of the theory from its historical basis, the "material" on which the general laws and factors worked. This division was certainly not new—it replicates the Historical School's own formulation, although with an emphasis on the opposite, nonhistorical, side of the dualism. But Weber's binary formulation is neither hierarchical nor successional. Instead, he implies that these two elements interact with one another, which would be the same tack taken by his cultural sociology.

In addition, the vitalism that he became taken with in Prague makes a brief appearance in his description of historical economic configurations. He describes these entities as "bodies" (*Körper*), and modifies them with adjectives such as "living" and "growing."[41] This organic language for the historical realm contrasted with that of the mechanistic sphere. These two elements—coexisting, interacting, yet essentially independent from one another—would be fundamental to his cultural sociology. The later civilizational-cultural dualism that employed this language is anticipated in the following passage:

> A...distinction that ought to be made is that of "natural-technical" and "social-cultural" factors of location. This distinction (also made in terms of the effects of the factors), which, however, cannot be fully made here..., can be signified, when it is made, thusly: The advantages that draw industries hither and thither may be given by nature. In that case they could be altered only by changes of these natural conditions, by the extent of the control of nature—in other words, by technical progress. They would be independent of the particular social and cultural circumstances; at least there would be no direct dependence. On the other hand, the advantages which draw industries hither and thither may be not natural-technical but social-cultural, the consequence of particular economic and social forms, certain levels of culture, etc. Then they would likewise depend completely on these things and their meaning.[42]

In Weber's later writings, the term "natural-technical" would be replaced by "civilizational,"[43] and "social" would become a separate entity, more akin to the civilizational than to the cultural. But the basic binary formulation of the cultural sociology is here.

A second connection between the theory of locations and the cultural sociology is even more inexact and is purely structural—the differentiation of elements that were "bound" (*gebunden*) from those that were "free." The first instance of this came at the end of *Theory of Location*, where Weber attempted to briefly develop a theory of the stages of the productive process, which involved the order of appearance of various "strata of locational distribution" (*Lagerungsschichten*).[44] An agricultural stratum (farmers) would be the first to appear, followed by a primary industrial stratum, which provided products directly for the former. This stratum, in turn, was provided with goods and services (for example, transportation) by a secondary industrial stratum. All three strata were oriented largely toward a specific location, which is why Weber did not distinguish local officials as a separate stratum. These three strata made up the economic "body" (*Körper*). The role of the remaining strata, termed central organizing and central dependent, was simply to strengthen the forms provided by these three. Of these latter two the central organizing stratum, which was described as "really independent," was the more important and only this will concern us here. It included "all those elements that are not concerned with local but rather with general organization and direction of the exchange of both material and nonmaterial goods".[45] This included officials with general organizing functions as well as intellectuals outside of the bureaucracy. Members of this stratum were "free" in their choice of location, so their locational distribution was something independent. When they were oriented to the "body," they were oriented to it as a whole. "Although the stratification of all these elements is diverse and subject to quite varying laws, they belong together in the sense that their stratification can be thought of only on the basis of the totality of the other previously named strata."[46]

Werner Sombart extended this distinction even further in his review of Weber's book, making the distinction between "bound locations" and "free" ones. Mines were bound to where ore deposits were located; firms using a type of labor found only in a certain place (for example, the high-quality workers of the big city) were bound to that place; firms that provided personal service were bound to the place of consumption. Weber's dynamic of location applied not to such bound firms, but only "where it was a matter of free locations." Weber adopted Sombart's terminology in the section of his later article that discussed the growth of capitalism from local markets to larger markets.[47]

What these passages have in common is the dualism of elements bound to limited positions and those that were free and hence attached to the system as a whole. Here, Weber was discussing economic roles, but when he developed his sociology of culture this structure would be transferred to the cultural realm. Members of political parties and interest groups (including the bureaucracy) were "bound" to certain locations in the sociocultural field. In this sense they were structurally similar to the primary and secondary strata of the locations theory. The intelligentsia, conversely, occupied the "free" location economically and a similar position with regard to the sphere of culture. A major difference would be that the intelligentsia would no longer be simply grouped with the bureaucracy as they were in these essays. Instead, the bureaucracy would be assigned, at least partially, to a separate sphere—that of civilization.

I noted above that these studies on the location of industries cannot be considered ideational studies. However, when Weber accepted a position in Heidelberg in 1907, he began to write ideational studies, including the initial formulation of the sociology of culture. Some of the structural elements of these new studies would be informed by the abstract economic theories he had just completed.

In this chapter, I have placed Weber within both camps in the "methodological controversy" in the economic sciences—first the Historical School and then the Austrian School. I argue that in both cases he included structural elements that moved him away from the formulation of the discursive coalition (even while being a part of it) and would translate directly into his cultural sociology. It is also the case that he did not forsake one of these schools for the other. He continued to publish in both areas, although his return to Germany brought him into closer contact with the issues of the Historical School, especially Social Policy. This relationship is also reflected in his cultural sociology. The civilizational sphere would not be demoted in importance beneath the cultural sphere, but it would be taken as a given, an essentially rational process that extended beyond any one time or any one nation. What would concern Weber the most was how the cultural movement of his time and his nation would respond to crises at least partially brought about by that process. In addressing these issues, his academic role changed from being that of a specialist to that of a diagnostician, and his diagnoses would bring him into direct disagreement with the discursive coalition.

3
Heidelberg and the Empire, 1907–1917

In chapter 1, I examined the discursive coalition that tied German professors to the imperial state, establishing their social status by giving them a share of the state's prestige. This social capital was augmented by the cultural capital that accrued from being recognized by the cultivated bourgeoisie and large sectors of the propertied bourgeoisie as the primary arbiters of the nation's cultural values. The academy was instrumental in shaping the discursive coalition's habitus, centered upon the belief that the nation formed an organic unity in which the divisive elements of civil society were subordinated to, and given meaning by, the hierarchically superior state. Throughout the empire this discursive coalition was able to ward off those elements that contested its hegemony. In the political field, these challengers were primarily those who believed in a parliamentary public sphere, whether the "public" remained a constructed unified one or a conflicted fragmented one. The challenge to cultural power came from a variety of groups: academic positivists (such as the Austrian School), socialist intellectuals, cultural pessimists, the experimental avant-garde, and adherents of vitalism (*Lebensphilosophie*), including those within the youth movement. Professors with close connections to the political establishment dominated the academic field. Others with ties to one or another of the oppositional forces, however qualified these ties were by the shared habitus of the professoriate, were consigned to more marginal positions. The latter included important members of Alfred Weber's generation within the Historical School of National Economy.

In his early years under Gustav Schmoller in Berlin, Weber produced a series of empirical economic articles that, despite some variations on the orthodox themes, were informed by the assumptions of the Historical School. That discipline assigned prime importance to the "social question": the use of state reforms to integrate the oppositional working classes into the organic unity of the nation. When he briefly left the cultural field of the German university to take a position in Prague, Weber experimented with the heretical abstract approach of the rival Austrian School. He returned to the German scene in 1907 to become a professor at the University of Heidelberg, where he remained until the Nazis forced him to retire in 1933. This chapter narrows the context of Weber's writings by examining three different lines of force of the discursive field within which these works were developed: the unique place of Heidelberg and its university; larger events, especially the world war and its challenge to the imperial power elite; and Weber's own career, which interacted with these other two.

Weber was originally filled with trepidation over leaving Prague for Heidelberg. Assuming the latter to be only a small provincial city in the Prussian-dominated empire, he feared it would not offer the kind of intellectual stimulation he enjoyed in Prague. On his arrival, he was surprised to find an extremely lively intellectual environment awaiting him. More than that, he entered a world that nurtured what I term his "monadic organic" view of the world.

The "Spirit" and "Myth" of Heidelberg

The institutional model already discussed can be supplemented with a geographical one. In the mid-nineteenth century, the "social politician" Wilhelm Heinrich Riehl offered a template for this model in his division of the lands of the German Confederation into northern, southern, and middle regions. He depicted the Prussian-dominated north and the south under Austria and Bavaria as the politically unified "centralized country." Separating these two regions, the middle lands between the Rhine, Danube, and Elbe rivers were designated as the "individualized country." The centralized country, in Riehl's eyes, was the more dynamic, Prussia and Austria having established themselves as great powers. The middle lands (including Baden and Württemberg), on the other hand, stagnated in their exaggerated provincialism, political fragmentation, and excess of cultivation. The individualized country, despite occasional spiritual awakenings, could not be taken as representative of the German people.[1] Riehl's pronouncement was seconded two decades later by the spokesmen of the empire's discursive coalition, now minus the Austrian part of the south and dominated by the north.

The inhabitants of the "individualized" middle had a very different perspective on Germany. It has been captured recently by Alon Confino's study of nationalism in imperial Württemberg.[2] Confino demonstrates how the particularism of middle Germany became identified with the empire in the minds of its small-town inhabitants. Nationalism succeeded because of the metaphorical identification of the local homeland (*Heimat*) with the larger national one. Local communities, through institutions such as Heimat museums and beautification societies, as well as "invented traditions" such as regional costumes (*Trachten*), imagined a national community that was embodied in regional uniqueness. Whatever drove this movement—often commercial endeavors—these efforts promoted the image of the nation as an organic unity of unique essences rather than a standardized whole.

I label this type of organicism as "monadic," because its subscribers viewed the "individuality," whether a person, an institution, a place, or an era, along the lines of the Leibnizian monad. Individual monads were unique and self-contained entities, not subject to meaningful subdivision. Their essential nature could not be exactly, or causally, explained or quantified. The same was true of their relationship with one another, as a preordained organic harmony, perceivable to some degree but not exactly demonstrable, existed among them. They reflected in a microcosmic way the underlying organic unity of the whole, usually described as an "idea," without losing their own individuality. I argue that this metaphor converged with a philosophy of culture that had a strong presence in the universities of the middle region. Its shining example was Heidelberg.

Heidelberg has been aptly dubbed a "world village,"[3] a local place whose cultural essence was that of the larger unified whole. As a university, Heidelberg was second in importance only to Berlin,[4] with whose surroundings it can be contrasted. In the empire and the succeeding republic, Berlin was the center of political power, and during the empire the university was firmly tied to that power. In addition, as the largest German city, Berlin was a center for social, political, and cultural opposition, from its militant proletarian neighborhoods to its broad modernist boulevards. Thus, the boundaries between the field of the discursive coalition and its environment were sharply defined in Berlin. Academics with ties to the environment were marginalized within the field. No one symbolized this better than Georg Simmel. As a Jew, whose fragmented sociology reflected the modernist city, who advocated vitalism and sympathized with the working classes, Simmel could never rise above the rank of lecturer, despite his significant publications and his popularity with students and the larger public.[5]

Heidelberg was an entirely different story. The town is located on the Neckar River, 11 miles from its juncture with the Rhine at the industrializing city of Mannheim. Its romantic setting complete with castle ruins, the strong presence of artisanal and agricultural influences, and the fact that everything was easily reachable by foot, made it something of a "greenhouse," sheltered from the outside world. When Alfred Weber arrived, the population of the town was about 40,000, with about 2000 students and 160 faculty at the university. It is alleged that every citizen of the town knew every professor by sight and name.[6] The fact that Heidelberg culture remained fairly consistent in the face of the sweeping changes that took place in Germany after 1918 reinforces the description of it as a sheltered greenhouse. The archeologist Ludwig Curtius claimed that only there did he feel "protected from the times" and that the rhythms of the Neckar described by the poet Hölderlin flowed in the writings of current Heidelbergers Max Weber and Heinrich Rickert. In the minds of its cultivated inhabitants, the city was characterized by its ties to nature and its easiness, vitality, and youthfulness.[7] At the same time, its intellectuals interacted with more cosmopolitan areas of Germany, creating a blend of cosmopolitanism and provinciality that became known in the memories of its participants as the "Spirit of Heidelberg." The law professor and later socialist official Gustav Radbruch described this "spirit" as completely unique,

> a unified spiritual world in which the spiritual people of Heidelberg moved, were influenced by it, and in turn influenced it. I do not believe that at any other time at any other German university an exchange of thinking by the various minds (*Geister*) took place to this degree. One has to think back to Jena in the classical age. There one found the same unceasing discussion, the same eternal conversation, the same *"Symphilosophein"* [philosophizing together], as it was called then. It also had the same active participation of clever and cultivated women in the spiritual world.... At that time Heidelberg was like a Noah's Ark in which a specimen of every new fashion among spiritual people was represented.[8]

These vital sources, the rich intellectual life infused with the "Spirit of Heidelberg," wrote Weber, were "for the newly arrived something along the lines of a revelation. It summoned history, philosophical existence, all of the old tradition before its tribunal."⁹ The "judges" in this tribunal were an "extremely unusual number of really important spiritual heads" who, whether academics or nonacademics, had ties to the university. In many ways the university dominated the cultural life of the town. Professors had great cultural capital reinforced by economic capital in the form of relatively high salaries.¹⁰ Robert Norton writes that:

> In Heidelberg, in particular, the life of a professor was not merely comfortable, it bordered on luxurious.... Extravagant dinners for a dozen guests or more were not uncommon, involving multiple courses accompanied by choice wines and often framed by a private musical recital or dramatic reading, all conducted within the spacious, tastefully decorated rooms of the professor's home and made possible by a permanent professional staff.¹¹

In addition to the established cultural elites, a very active group of students participated in the intellectual life of the town and looked to their teachers as intellectual role models and not simply as instructors. Marianne Weber described this atmosphere:

> In this situation the view developed that the universities, as centers of spiritual life as well as educational institutions, must not content themselves with imparting knowledge to the young generation and sharpening their spiritual tools, but that they had additional tasks: forming the total personality, imparting convictions and ways of thinking, instilling a practical, evaluative stand on all great problems of life, reconstructing an undivided conception of the world, and making ideological proclamations. These tasks did not only concern theology and philosophy. The other cultural sciences seemed to offer a wealth of opportunities as well, and the social sciences and history were primarily charged with cultivation of *political* will. Against this background of evaluating consciousness and a world-view—which one had not been decided—the fragmented scholarly disciplines would again join forces and be united. That is why, so people thought, a student should find in his university lecturer not only a teacher but also a *leader* who would set goals for his will and guide his personal development in the right direction.¹²

Despite most professors' emphasis on specialized scholarship over larger goals of cultivation, students looked to the university for more than scientific training, seeking the development of the personality and the reconstruction of a unified picture of the world.[13]

These students were often part of the youth movement that had grown since the beginning of the century. That movement, infused with vitalism, was critical of the materialist society of the empire, believing that it prevented people from becoming integrated human beings.[14] Many believed that the need for the wholeness of the personality and its integration into "life" was not being fulfilled in their university classes. In 1913, Georg Simmel warned that if the German universities did not address these needs, then students would seek answers in more mystical movements. The university, he lamented, had given up the "inner leadership" of youth.[15]

Heidelberg was the exception to the rule. While students might not often find guidance for life in the classroom, strong ties existed between some professors in the human sciences and their students, beyond the university's walls. This "spiritual sociability" allowed the continuation of the notion of a community of scholars at a time when it was declining at other universities.[16] Alfred Weber dedicated himself to this notion of community.

Of special importance in fostering this atmosphere were the salon-type "circles," where a wide range of issues were discussed. These circles were aimed especially at the younger generation and, while having various orientations, did not remain isolated from one another. Their interaction and the rich variety of intellectuals, including socialists, vitalists, feminists, Russian and Hungarian emigrés—a spectrum of academic and political actors—provided a cosmopolitan energy.[17] Despite disagreement among the participants in this milieu, what excited Weber was the "strong and productive exchange of all its forces." This "eternal conversation" in professors' houses and cafés fueled the optimistic belief that some kind of synthesis could be shaped from the diversity of the new sociopolitical commitments.[18] This tolerance of intellectual diversity on the part of the state of Baden and the cultivated public of Heidelberg gave the university the reputation of being the most liberal in Germany. This liberalism accompanied a belief in the existence of a public that was dominated by the cultural and political elites. There was less willingness to recognize that the old unified public, perceived as more or less organic, had given way in Germany to a fragmented one that did not give due deference to the cultural elites.[19]

The liberal orientation of many professors and their interaction with intellectual forces beyond the university meant that the boundaries between the discursive coalition and the environment of opposition that were so sharply defined in Berlin, thanks to its industrial and modernist physiognomy, were much more muted and permeable in the cozy atmosphere of Heidelberg. One could have an earnest debate with a socialist intellectual without the accompanying threat of large proletarian districts crammed with tenements. Ludwig Curtius reflected these conditions when he wrote that there was "a spiritual fluidity in the air" that fostered the goal of "a cultural-scientific synthesis of the European and German spirit with the promotion of a new ethical-moral-political bearing of the individual and ultimately the nation."[20]

Weber's junior colleague Karl Mannheim wrote that this sheltered "atmosphere" was, in a way, representative of all Germany, for the national culture was decentralized through its emergence in towns like Heidelberg. These German towns were not "provincial" in the sense of being on the periphery of cultural development, but rather formed multiple "vital sources" of that development. The perceived embodiment of the national cultural whole in the multiple individual communities (individualities) reinforced the monadic organicism that will be discussed further in the next chapter.[21] Those thinkers who subscribed to this organicism tended to view the larger unity as culture and to emphasize the correspondence between the culture and the personalities engaged in its creation as well as those who received it through formative institutions such as the universities. This view could be found in varieties of both German historicism and the ideal of cultivation, which were akin to forms of vitalism, from their common eighteenth-century founding figure Wilhelm von Humboldt to Weber's Heidelberg colleague Ernst Troeltsch and the Leipzig pedagogue Eduard Spranger.[22]

The individuality of a person was often represented as the soul (*Seele*), an intrinsic life force that developed internally rather than according to external sources. The soul was presumed to possess an inner unity that preceded acts of feeling, willing, and thinking. The interaction of this primordial personal unity with other individualities took place in the realm of spirit (*Geist*), where it was conceptualized and could be objectified. This potential for objectification distinguished spirit from soul by virtue of its quality as a communicative medium. It must be noted that none of these terms were applied consistently. There were certainly inconsistencies in Weber's own writings.[23]

This monadic organicism, fed by the Spirit of Heidelberg, led a number of the faculty, including Alfred Weber, to see Heidelberg as a model for all of Germany. As one of them stated:

> Precisely our Heidelberg, which takes care to admit among its matriculants names from every region of Germany, is called upon primarily to solidify and strengthen our academic youth in the spirit of German unity. Those who are employed and teach here serve the entire German fatherland.[24]

The belief that the Spirit of Heidelberg reached beyond the town to all of Germany has been termed the "Myth of Heidelberg."[25] As with all mythic formulations, reality was smoothed of its rough edge to fit the ideal.

Mannheim was aware of two important shortcomings of this mythic presentation of culture. First, a gap existed between the native intellectuals and the citizens they encountered in everyday life, so that they were closer to a foreign thinker like Mannheim himself than they were to their fellow citizens. The deferential attitude of most town residents toward the professoriate gave a deceiving impression of unity. Thus, Mannheim saw the intellectual as a part of the Heidelberg culture but, at the same time, distanced from the everyday core of that culture.[26] Second, despite Weber's own continued involvement in national politics, the Spirit of Heidelberg was never more than a minor player on the larger political stage. This fact doomed Weber's attempt to establish a closer relationship between the cultural elites he represented and the main political actors, be they the autocrats of the empire or the democrats of the republic. He was unable to successfully establish either an oppositional discursive coalition during the empire or a new one in the republic.

Weber's prewar years in Heidelberg

Volker Kruse correctly argues that the six years immediately after Weber's 1907 arrival in Heidelberg were formative for his career.[27] During this period, he began to reformulate the "social question" as the "cultural question," which meant moving away from the empirical studies of the Social Policy Association to something much more speculative and inexact, a version of the aforementioned "ideational studies." The distance that the years in Prague had already created between him and Schmoller was increased with Weber's return to Germany. However, he did not move in the direction of his abstract and geometrical *Location of Industries* because that approach did not address the issues of orientation and values that were central to the perceived cultural crisis. Rather, he moved toward vitalism, which had become increasingly popular in Germany after 1890.

Weber's cultural concerns can be divided into two categories, culture proper and its application to politics, each of which can, in turn, be further subdivided into two categories. His cultural writings attempted either to define the cultural crisis he believed to exist or to delineate a cultural sociology necessary to analyze that crisis. His political writings applied his cultural positions to the distribution of power within Germany and to Germany's role in the world, that is, to domestic and foreign policy. While these four areas are presented here as distinct, they were not separated in Weber's writings. A belief in the centrality of culture defined Weber's world view for the rest of his life, although there would be changes in its explication within the cultural-political fields in which he wrote.

In defining the role of culture and its crisis in the years before the war, Weber had to come to terms with the "eternal conversation" in Heidelberg, and, specifically, the salon-type "circles" of two mythic figures, his brother Max and the poet Stefan George, both of whom had only marginal ties to the university.[28] Despite his interaction with participants in both of these circles, he did not adopt the positions of either of them. This was especially problematic with regard to his brother.

Alfred's ambivalent relationship to Max took several forms. The fraternal rivalry, always there, was undoubtedly exacerbated by Max's emergence from mental illness and their closer geographical proximity.[29] Personally, the two competed for the affection of Else Jaffé, the wife of Max's coeditor, Edgar Jaffé, a competition won by Alfred.[30] Politically, the two brothers remained close allies within the Historical School and the Social Policy Association, where both were leaders of the rebellious third generation.[31] Alfred was an important participant in debates at the Association's general meetings,[32] along with having other organizational and editorial responsibilities.[33]

Although both brothers believed that the old discursive coalition was no longer viable, they differed on what would succeed it. Max formulated a type of rational individualism borne of his interest in ascetic Protestantism and his recent trip to America. He disapproved of Alfred's enthusiasm for vitalism and his resulting interest in eroticism and sexuality—themes that were very much in vogue among Heidelberg intellectuals.[34] Although Max depicted Western modernity, with its pillars of capitalism, bureaucracy, and science, as having become a sterile "iron cage"[35] characterized by value pluralism, he dismissed any hope that some kind of alternate organic community could be realized. Nor did he believe that science could solve these problems. Instead he placed his hopes on the responsible political actor operating within modern democracy. Valuation was arbitrated by political decision-making, not by academic science. Nor could science legitimize such political activity. Its function was essentially instrumental: to present the consequences of various courses of action available to the political actor without choosing among them. Max articulated the politician's "ethic of responsibility" and the limited handmaiden role for the scientist in two lectures given at the juncture of the imperial and republican periods, "Politics as a Vocation" and "Science as a Vocation."[36]

At the other pole from Max stood the poet Stefan George, who largely rejected modernity, and especially modern mass culture. His romantic anticapitalism called for "no interest, no machines, no human masses." America, an inspiration for Max Weber, was condemned by George as the antithesis of everything he stood for. He offered his followers what amounted to a religious cult, a closed and unified world-view presided over by a charismatic master and organized as an organic community removed from modern society. Vitalism played an integral part in this world-view, and, accordingly, modern science and university "intellectualism" were denigrated with the dictum "No road leads from me to science." He, like Max Weber, advocated the political over the scientific as the source of meaning and values, even if his politics—aesthetic and aristocratic—was diametrically opposed to Max's.[37] The competition between these two circles was intense, giving rise to heated discussions. It continued even into the republic, when Erich von Kahler, a member of the George circle, published an attack on Weber's "Science as a Vocation" from a vitalistic perspective.[38]

Alfred Weber sought to mediate between the competing poles represented by his brother and George. If Max's position was too rational and ascetic for Alfred, he also rejected the opposite extreme represented by George. Although he admired George's poetry (as did Max) and had considerable contact with the poet's circle, especially with Friedrich Gundolf, he believed that the group was characterized by an escapist, irresponsible self-absorption, which guaranteed its isolation and inability to grapple effectively with the important issues of the day. He thought that the wider influence of the group, especially among the youth, was a "colossal danger," for all they produced were "soap bubbles without content."[39]

Alfred never accepted the demarcation of science from politics that both his brother and George made.[40] His desire to create a new discursive coalition can be seen in his range of activities. In addition to his political engagement and university duties, he formed the Janus Circle, which was attended by important figures in Heidelberg society, many of whom also attended the more renowned circles of Max Weber and George.[41] In keeping with the "Spirit of Heidelberg" that these circles fostered, he presided over "Sociological Discussion Evenings," in which professors and students met weekly or biweekly, often to discuss recent publications. Students played a very important role in this group.[42] This emphasis on interacting with youth was also evidenced by Alfred's strong connections to the German youth movement. At the famous gathering on the *Hohe Meißner* in 1913, he was one of the professors who addressed assembled youth groups that later became the Free German Youth.[43]

Youth were not the only element of the larger cultivated public with whom Weber had ties. He had some contact with the German *Werkbund*, an association of artists and industrialists founded in 1907 for the purpose of preserving German culture in a materialist age.[44] And the location of his publications changed. Whereas he submitted most of his earlier writings to journals for specialists in Social Policy and economics, now a large number of his publications appeared in lay forums such as *Die neue Rundschau* and *Die Tat*, which had a wide readership within the cultivated public sphere. These activities represent a major shift in emphasis among the roles outlined by Kruse from specialist to diagnostician.

As noted above, what many of these groups and activities had in common was an openness to, if not outright acceptance of, vitalism (*Lebensphilosophie*). Weber's attraction to the works of the French vitalist Henri Bergson, which had begun in Prague, increased after his return to Heidelberg, in part because of his circle of acquaintances. He developed ties to the popular philosopher Hermann Keyserling,[45] a friend of Bergson who also had close ties to Wilhelm Dilthey and Georg Simmel, both important exponents of vitalism. Two former students of Simmel, Ernst Bloch and Georg Lukács, attended gatherings at Weber's house, as did Friedrich Gundolf of the George circle. Hans Driesch, a neovitalist biologist, was also an important acquaintance.[46] Vitalism, which formed an important influence on Weber's activities and his writings throughout his Heidelberg career, will be discussed in the next chapter.

The world war

At the beginning of the war, the three major themes of Weber's writings—culture, domestic politics, and foreign policy—came together dramatically among most academics in a sense of community labeled the "Ideas of 1914." With the votes for war credits by the socialists, many observers believed that the fragmentation of imperial society along class lines had been overcome and that Germany had once again become a spiritual, that is, cultural, "national community."[47] The alleged purpose of this coalescing was the defense of German cultural values against the shallow materialism of Britain and the brutal autocracy of Russia.

In 1914, German professors, including many at Heidelberg, sought to strengthen morale on the home front and to challenge the successful British propaganda among neutral nations.[48] These German writings had two major themes. First, they attacked the idea that the autocratic German state was reactionary, to distinguish it from Russian autocracy, arguing that it actually protected people better than did the liberal parliamentary British state. This theme, then, focused on the domestic state-society relationship and represented a reiteration of the imperial discursive coalition. Second, they disputed the British claim that the German state was the aggressor in the war, arguing that Britain bore primary responsibility.

German academics portrayed the German state as more peaceful, more progressive, and more concerned for the welfare of its people than the British state, because in Britain amoral civil society held a position superior to the state. These writings took the vertically arranged dualism of the discursive coalition and presented it as a truly binary opposition. Instead of two elements, one subordinate to the other, it was now a matter of two conflicting forces, one of which was positive, the other negative. Instead of two aspects of the same culture, the dichotomy was now strongly presented as two separate cultures combating one another. In short, civil society of the discursive coalition's formulation was projected onto Britain. This amounted to an extension of the derogatory term that had been used to describe a civil society without the integration provided by the state—"Manchesterism." During the war the dichotomy reached its most simplistic level when Werner Sombart portrayed it as the British materialistic *Gesellschaft* versus the German cultural *Gemeinschaft*, or "traders" (*Händler*) versus "heroes" (*Helden*).[49] The most famous expression of this view was the "Appeal to the World of Culture," on October 4, 1914, by 93 intellectuals, including many professors. They denied Germany's guilt for the war and declared that Germany would fight to the end "as a cultural people."[50]

Despite sharing this common defensive position at the beginning of the war, academics quickly divided into two camps, "radicals" and "moderates." The radicals supported annexations of territory and reparations payments from the Allies, and they resisted any reform of the imperial government. The moderates, conversely, maintained their defensive position, eventually advocating a negotiated peace. Although a number of moderates advocated reforms of the political system late in the war, most rejected the replacement of the monarchy by a republic.[51]

This essentially defensive attitude of the moderate German academics toward their state led to their claim that British foreign policy, not German, led to World War I.[52] They argued that it was false to portray Britain as the defender of the balance of power against German *Weltpolitik*. Instead, they said, one must look at the larger world situation. There, it became clear that Britain, with its huge overseas empire, was the dominant world power. Maintenance of the balance of power on the European continent was a means to Britain's preservation of its world mastery and its dominance of international trade. German policy in the nineteenth century was simply aimed at securing Germany "a place in the sun," not at acquiring any kind of world mastery. In this sense, they argued, Germany sought to establish a world-wide balance of power. Thus, what the British represented as an attack on the concept of the balance of power was actually a defense of the concept from the German perspective. It is important to note that these thinkers did not attack the concept of the balance of power, which had a defensive ring to it, but rather the British formulation of that concept. This went along with their general view that the war was a defensive struggle to preserve their growing economy against British aggression, as well as to preserve the Austrian Empire against Russian aggression. In Heidelberg, the moderates were the strongest contingent, and its faculty senate rejected annexationist petitions from other universities.[53]

In responding to the wartime events, Alfred Weber adjusted his policy arguments several times, but he maintained his basic orientation. He never abandoned questions about the nature of the German state and its relationship to society. Those questions involved a good deal more foreign policy and were asked from a cultural perspective grounded in vitalism. His monadic cultural presuppositions were in accord with the views of many moderate academics.

The coming of the war delayed Weber's attempts to develop a specific cultural sociology.[54] In fact, during the war period, the number of his publications declined. His most important work during the period was a selection from his letters to Else Jaffé, which appeared first in *Die neue Rundschau* and then as a book.[55] Like a great many other academics, he was caught up in the initial enthusiasm over the war, believing in the "Ideas of 1914," the creation of a new community that had overcome class divisions. Although 46 years old, he volunteered for the army and served on the Alsacian front for a short time. In 1916, he was transferred to the Ministry of the Treasury in Berlin, and, three months later, was discharged. Even while in the army he was active in politics, helping to organize two conferences on war aims. That activity increased after his discharge.[56]

Weber's wartime letters focused upon two themes that would be central to all of his wartime writings—Germany's place in the world relative to other European powers and the configuration of the domestic power structure; or, for what was Germany fighting, and who should be leading the fight? Both questions had a cultural premise, because Weber assumed that Germany was fighting not for actual territory but for cultural supremacy in Europe. He also advocated an important role for cultural leaders in German decision-making. In both issues, he made "spirit" (*Geist*) a central term, whether in Germany's "spiritual world mission" or the "significance of spiritual leaders."[57] And, in both cases, Weber's argument was structured by the same monadic organicism that characterized his cultural writings. It is to those readings that we must now turn.

4
The Question and Sociology of Culture

As Alfred Weber's generation challenged their predecessors within the Historical School of National Economy and moved toward "ideational studies," they looked increasingly to culture as an alternative to the centrality of the state. In 1909, Weber began his cultural turn. Although he continued to publish short articles on economics, they were secondary to his interest in culture and its application to politics. Rather than use cultural concepts to examine capitalism, as his brother Max and Werner Sombart did, he sought to use it to reshape the classical ideal of cultivation (*Bildung*). He rejected the strong connection of the ideal of cultivation to the state in favor of a more purely cultural version of the ideal that retained its organic assumptions. Weber's view of culture, which he constructed around the concept of the monad presented in Chapter 3, was reinforced by the setting of Heidelberg and by his attraction to vitalism.

This chapter will address two of the three components of Weber's "cultural question": the definition of its role as the locus of the organic unity of the nation, and the outline of a discipline to examine more closely the elements of that role—the sociology of culture. The third component, culture's application to the political realm, will be the focus of the next chapter.

The initial cultural writings

In the Winter Semester of 1909–1910, Weber scheduled a lecture course, "Cultural Problems in the Era of Capitalism." Due to his additional administrative duties, the course was postponed until a year later and then repeated the following year.[1] Its title, which characterized the era in terms of capitalism rather than political institutions, was in accord with other works of his generation, but it was a grand cultural-historical survey that emphasized philosophy and art more than economic structures. It also included some programmatic statements that would be reiterated in his essays on the sociology of culture. The course was never developed into an extended publication during the empire. Instead, during the next five years he published a series of short pieces in a variety of forums: not only his brother's academic journal and the proceedings of the German Sociological Association, but also publications of the German youth movement, journals that were aimed at the larger cultivated public, and newspapers.

The first of these was a short essay, "The Cultural Type and Its Transformation," published in a student journal. Here, Weber took a position at odds with that of the discursive coalition with regard to culture and cultivation, by detaching them from the bureaucratic state. He wrote:

> I do not know: if the cultivated person (*Gebildete*) of our times still has a feeling that he is, and in what way he is, the "fashioner" (*Bildner*) of the cultural model of his times; if there still lives within him something of the ecstasy and the elevated consciousness with which his "ancestor in history," the person of Renaissance culture, pursued "self-fashioning" (*Sich-Bilden*) in the sense of consciously giving the self a form and in so doing creating a model for others; if a glimmer still lives in him of the self-perception of the "person as work of art," that most beautiful relic of the Renaissance that [Jacob] Burckhardt again drew out for us; or if utility (more specifically the banal function of acquisition by which so-called cultivation has been degraded since the modern bureaucratic state made it the prerequisite for "vocation" and "position")...has not obstructed the view of the heights of really becoming cultivated. I do not know how it stands with this, although I fear the worst for this utilitarian Germany of today with its pronounced instinct for spiritual small change.... The fact is that the manner in which the cultivated man "sees himself," the conception that he possesses of that out of which he has to configure himself, his cultural consciousness, has the same fundamental significance for the culture of a people today as it had earlier. It creates the standard of the cultural mean of the people...not through pure imitation but through conscious means, such as the press and other things; and it creates this standard in completely different organizations than before. But the cultural consciousness that the cultivated man has of himself is in fact, today as before, the

basic element for all cultural development of the masses. It is the soil into which the new ideas and objectifications sink and from which they must be nourished if they are to maintain a living significance for their time.[2]

In claiming that the cultivated elite established meaning not only for themselves but also for the entire nation, Weber was very much in accord with the ideal of the discursive coalition. Cultivation was important to both the modern state and the nation's cultural consciousness. However, Weber departed from the discursive coalition by presenting these forces not as complementary but as antagonistic. The state had "degraded" cultivation by turning it into merely a prerequisite for position, a means to a material goal.[3] Cultivation had been diverted from its role in the cultural mission that typified the Renaissance man. The juxtaposition of culture to the bureaucratic state, a clear departure from the discursive coalition, would characterize Weber's writings over the following years. This dualism coincided with his primary one of civilization and culture, which was central to his cultural sociology. In this schema, the state was assigned to the civilizational sphere.

Weber's use here of the Renaissance, which enjoyed cult-like status both before and after World War I, is instructive, as is his choice of Jacob Burckhardt. In art, literature, and the human sciences, the Renaissance was often called upon as a trope for cultural individualism and a lament about nineteenth-century materialism and its naive optimism. Burckhardt emphasized the precarious connections of individual creativity and spontaneity to larger cultural entities. His emphasis on the individual within a cultural cross-section of history was in accord with Weber's own view of culture. Alfred's approach contrasted with that of his brother Max, who emphasized the Reformation over the Renaissance and tied it to the beginnings of modern economic individualism.

Weber portrayed the nineteenth century as the era when the monadic cultural mission of the Renaissance man declined to its present state. Early in the century, depicted as a golden age of Idealism, creative individuals had "a capability of self analysis, a possibility to discover worlds and the contrasts of worlds in our inner being, and with this to bring the entire cultural contents of human history into connection."[4] This great age of the cultivated person disappeared as the century went on, and the multidimensionality of the creative personality became "crusted" over. The "Romantic" thinker was replaced by a person who loved prosperity "but was not suited to solve a single one of our great cultural questions, because he could no longer feel them in their core." Weber used Theodor Mommsen's term "rebarbarized" to characterize this new man, whose concern with utility restricted his cultural awareness.[5] Accordingly, the imperial state and its officials were portrayed not as the creators of German greatness, as Schmoller claimed, but as contributors to the process of cultural decline.

The loss of these ties between the monadic creative individual and the larger organic culture comprised the crisis of culture as depicted by Weber. But before discussing his analysis of that crisis and his hopes for its solution, it is necessary to make a brief excursus and examine the movement that had such an influence on his formulation: vitalism (*Lebensphilosophie*).

Vitalism

In 1910, Weber sent Else Jaffé a copy of Henri Bergson's *Introduction to Metaphysics*, "so you can see where my intoxification with individuality originated."[6] The next year, he returned to Prague, where he had been initially introduced to Bergson's vitalism, to give a lecture on "Religion and Culture" that incorporated vitalistic elements. In that lecture, he did not attempt to biologize culture, but rather he adopted structural elements from Bergson's philosophy to apply to the cultural sphere. These would be supplemented by structural elements from the biology of Hans Driesch, who became Weber's colleague in Heidelberg.[7]

German vitalism had its roots in the late eighteenth century and was influential in the construction of the ideal of cultivation that Weber felt was being degraded by its attachment to bureaucratic training. Many of these thinkers and later vitalists used language similar to that used by the Romantics at the beginning of the nineteenth century.[8] Later vitalism, such as that of Bergson and Driesch, was most concerned with "whether organic processes are reducible to the same kind of mechanistic laws that govern physics and chemistry."[9] This section will examine the work of those two thinkers, who were important for Weber's formulation of culture.

Bergson depicted reality as "life," which he defined as a vital impulse (*élan vital*), a fluid stream, a duration. The flow of life had no linear direction, but rather spread outward from its center, oscillating on its outer circumference. He used the metaphor of a shell bursting into fragments, which in turn burst into more fragments. The particular way in which a shell bursts is determined by two factors: the explosive force it contains and the resistance of its metal casing. The explosive force is the vital impulse of life, and inert matter is the casing that restricts life until it explodes and that affects the pattern of explosion. Evolution is life escaping and shattering form. Bergson described this vital content as continual creation, as always in the making, as tendency. Reality is not being but becoming.[10]

Bergson used the paradoxical term "multiple unity" to describe the duration of reality. He meant that although the world appears as multiple forms, the reality flowing through and exploding those forms is unified and absolute. There are no discrete parts in this duration, and thus there can be no juxtaposition, no contradictory positions, no relativism. Just as the whole of a personality is contained in every psychical state, so the unity of duration is contained in every manifestation. To grasp this means to penetrate the multiple states to the unity behind them.[11]

Hans Driesch, Weber's colleague in Heidelberg, offered a somewhat different version of vitalism. Like Bergson, Driesch believed that life is a unified continuum that takes multiple forms. But Driesch denied that one can penetrate the forms to apprehend this unity. One can know life only through its individual manifestations.[12]

The difference between the two men is captured by their central concepts: Bergson's "vital force" and Driesch's "entelechy." On the one hand, the former is content, a force whose explosive power and the resistance to it by inert forms shapes the world. Entelechy, on the other hand, is not content but form, vital form. Like Bergson, Driesch wrote that form restricts and releases energy.[13] But for him, this process of restricting all possibilities and then allowing only certain ones to be actualized is the creative aspect of life and not simply an inert restraint to be exploded. While Bergson emphasized the unity of duration, motion (content) that resists rest (form), Driesch emphasized the manifoldness of actuality, rest that restrains motion.[14] While Bergson distinguished between vital content and inert form, Driesch largely ignored questions of content and distinguished between vital form and inert, mechanical form.

Here, Driesch distinguished between the singular causality of the inorganic world and the unifying, or individualizing, causality of organisms.[15] For example, taking a knife and cutting an earthworm in half is an example of simple causality (of an external force upon the worm). When the earthworm regenerates the missing half, this is an example of unifying causality (of an internal force exerting itself). This internal, unifying causality, the presence of an ideal whole, a hereditary potential (*Anlage*), characterizes entelechy.

Every organism has a preprogrammed "constellation," a form that it eventually assumes, despite the relative independence of the single events leading to it.[16] Driesch distinguished the constellation from the structure of the machine, which consists of a mechanical relationship of parts. When one removes parts, the machine ceases to be that machine and becomes something else. The same is true for chemical reactions; when hydrogen is removed from water, the latter ceases to exist. But despite losing parts, the earthworm remains an earthworm. While the parts can be removed by external causes, its constellational unity cannot.

Both Bergson and Driesch drew analogies between organic life and the human soul (*Seele*). While Bergson emphasized the durational content of the soul, Driesch emphasized its entelechy, its form.[17] This different emphasis led to two very different methodologies. Driesch, who began his career as a biologist, relied on experiment and logic to formulate his theories. Bergson, on the other hand, looked beyond the intellect to nonrational faculties in order to grasp life in its unity.

Bergson juxtaposed instinct and intelligence, the latter distinguishing humanity from other life forms. He defined instinct as "sympathy," being put in touch with the intention of life. In humans, instinct becomes "disinterested, self-conscious, capable of reflecting upon its object and of enlarging it indefinitely"—it becomes "intuition." Intuition, then, is intelligence that establishes sympathetic communication with the rest of life, with the absolute unity of duration. Intuition is distinguished from a second type of intelligence, intellect, or analysis, which is oriented to the inert, the purely material.[18]

Analytical knowledge, Bergson held, is practical knowledge. "We are geometricians only because we are artisans." One must approach an object geometrically, that is, logically, in order to utilize it under certain circumstances. Intellect represents the power to decompose according to laws and reconstitute according to systems, calculating with concepts such as causality and mechanism. It views matter as cloth to be cut out and sewn up. It "is the faculty of manufacturing artificial objects, especially tools to make tools." The "progress" of science and reason results from the intellect's increasing adeptness at making these concepts clear as it puts them to use, "clearness" meaning the certainty of manipulating a concept profitably.[19]

The very attributes that allow analytical intellect to achieve practical mastery of the world prohibit it from gaining insight into the duration of reality. To penetrate to the absolute reality, the mind must "reverse the direction of the operation by which it habitually thinks," suspend conceptual analysis, and place itself inside of mobile reality through intuition. But one can never begin at the level of the conceptual. "From intuition one can pass to analysis, but not from analysis to intuition."[20]

Weber's cultural writings borrowed from the work of these men.[21] Bergson would be more important to his separation of human activity into the two spheres of civilization and culture, but Driesch would be more important to his understanding of the nature of the cultural sphere.

The question of culture

Weber's vitalism appeared in his Prague lecture on "Religion and Culture." He adopted vitalist principles in defining the central issue of what he perceived as a cultural crisis: the need to establish a unified sense of meaning in the increasingly mechanized modern world. "To become cultural we must...comprehend analogous elements as belonging together and configure (*gestalten*) them."[22] This configuration required the location of a cultural center that allowed for a diversity of expressions but which would free people from the inner contradictions of the world by lifting life forces out of their "mechanical bracketings" and creating a coherent meaning.

Weber believed that while religion had often served as the cultural center, it is simply a part of culture and not the basis of culture. In fact, other elements of culture are better equipped to provide this center in the modern world. Religion best provides meaning in ages that are not characterized by rapid change, when existence is seen as "being" rather than "becoming." But since the Enlightenment, the thinkers have learned the importance of causal analyses in order to clarify relationships in a world that is no longer static. To ignore such a dynamic approach, to avoid meaningful interpretations of historical change as religions do, is to deny life in the present and retreat to the past.[23]

However, the very "apparatuses" that remove religion from its central position in today's world are themselves not the solution. In fact, these apparatuses, which are three in number, while necessary, could present a great danger to the establishment of cultural meaning, should they become too dominant. The first consists of intellectual forms, representations, and concepts, such as scientific laws. The second is the external means by which the first is turned into praxis. The rationalization of life becomes an external life-apparatus in which means predominate over meaning. This would seem to include things such as technology, business practices, and bureaucracy. Weber used his brother's example of inner-worldly asceticism producing capitalism. These two apparatuses can never be destroyed, and moreover, they are essential for existence in the world. (Both of them Weber would elsewhere assign to the civilizational sphere.) However, they must be qualified by cultural forces that preserve both the unity and diversity of life.[24] In short, forces from the civilizational process cannot in themselves organize the cultural sphere.

This is especially the case with the third intellectual apparatus, employed by rationalism—the establishment of absolute ethical and ideal norms and their application to inner life. The strongest and most important of the three, it is also the most dangerous in that it allows for no diversity. All living forms are forced to yield to these mechanistic norms. Weber called for its replacement, for while rationalism can provide mastery over the external world, it cannot provide meaning and morality for inner life. "The living (*Lebengiges*) will take its place."[25]

Here, one can see the vitalist influence on Weber's thought. The idea that rationalist concepts stem from the necessity to utilize the physical environment, but that they are limited in their ability to comprehend deeper life forces, is a central vitalist point that he borrowed from vitalist biology:

> We are released from these [purely causal] perspectives and from this rationalism only in that we no longer view this entire manner of intellectual comprehension of existence as the final, the only one and above all (this is perhaps the greatest and epoch making service of Bergson) as that which can provide and unveil the deepest of its inner contexts for us.[26]

Weber's distinction between the civilizational apparatuses of rationalization and cultural forces that put one more closely in touch with the living would persist throughout his career. One also sees here the distinction between the universality of civilization and the heterogeneity of culture. As it was for the vitalists, life (culture) for Weber is both unified and manifold at the same time. This represents monadic organicism with new underpinnings. In a 1912 letter to Hermann Keyserling, Weber wrote that when he "had the courage...to think to a certain degree monadologically" he replaced forms of thought such as the series with the "expression" in order to approach the living.[27] In his lecture he wrote:

> It is the only form according to which the living can be configured. And as in our creation of organic conduct from the growing center of the person, so the principle for our inner articulation and our merging into life will be that of the inner formation that has sprung from its own laws and the *similar* formation of similar parts of life that have the same inner relationship.[28]

This concern with shaping a new cultural person dominated his writings during this period, whether it was his attention to the youth movement or the postulation of a new Social Policy for the working classes. This new person would have to establish for himself or herself the kind of meaning once provided by religion.[29] He understood the precariousness of his position "swinging back and forth half-starved between stagnating schematization and directionless empirical relativism."[30] He nevertheless called on his audience to develop a religious-like yearning for it.[31]

Any viable culture, from Weber's perspective, has to provide a meaningful unity for two important sets of divisions: those of generation and those of class. In discussing the youth movement, Weber examined what would be involved in "creating who could be the bearer in its unity of a coming culture" among the next generation.[32] He found the current school system lacking, because it was unable to tie the narrow biological community of the family to a larger spiritual "body." It had simply continued this narrowness by preparing students for a profession through specialization, the result being a narrow egoism that could not make the passage from the narrower to a broader community. Weber saw hope for a new "spiritual type of education"—similar to that of German Idealism but which sought an "objective spirit" other than the religious one of that earlier age—in experiments currently underway, especially those of Gustav Wynecken, the leader of the Free German Youth. Only such communal schools could reintroduce universal education, instead of the existing specialized disciplines, to develop a new culture and will. Using vitalist language he declared:

> In the *cultural* sense, personality is not an end in itself but only function, only the participation in the life of the spirit; and that to which we must educate the individual is this participation. It is the psychic formation that allows him to take part in the *total*

consciousness of humanity through inner lived experience and immerses him in this total consciousness so that he feels himself to be an organ of this spirit.[33]

In a February 1914 speech to the Free German Youth, Weber reiterated the necessity to establish a center from within. "Above all they [the youth] have the primary need to realize themselves, above all to find a center, some kind of middle for themselves, and only from there operate outward and configure."[34]

He said that his generation placed priority on the relationships of the material world, such as the role of the state, the organization of the economy and the inclusion of the proletariat in the nation. Faced with these major social and political problems, they had no alternative but to live "from the outside in." But the new generation, especially those who are in the youth movement, no longer faced with so many major external issues, had "a tendency to live from the inside out." Such an endeavor, which required a new inner being and responsibility, could not be accomplished through already fixed forms, including the existing schools. Educational authorities should not try to regiment youth, for that would be counterproductive. On the other hand, Weber rejected the "fool's freedom" of simply trying to escape from the world and do whatever one liked. Freedom required preparation and discipline but from within. Here the older generation could be of help, not by dictating but by assisting in responsible individual self-development.[35] Here, again, one sees the contrast between the civilizational state institution, the school, and more the fruitful cultural entity, the monadically heterogeneous youth movement.

Weber expressed the same concern for individual self-development in a polite disagreement with his student Hans Staudinger in the preface to the latter's book, which was the first in Weber's series on cultural sociology. The arguments were later repeated in the periodical *Die Tat*.[36] Staudinger, a "committed Social Democrat,"[37] asserted the superiority of working class associations to those of the bourgeoisie, and in so doing used Tönnies's typology of community (*Gemeinschaft*) and society (*Gesellschaft*). The medieval association, he wrote, took the form of a "fellowship" (*Genossenschaft*), in which like individuals had organic ties to the association and "lived only through the *Gemeinschaft*." The rise of rational individualism in bourgeois society had destroyed this organic unity. While this has initially produced some impressive cultural creations, it ultimately resulted in a materialistic individualism that is culturally sterile. The association of bourgeois society had become a *Gesellschaft*.[38]

Staudinger believed that the working class association had again become a *Gemeinschaft*. The world of the worker demanded "concretism," not abstract thinking. In this world, workers developed close ties to one another, especially when they were "party comrades." In worker corporations, the individual was subordinated to the collective will, but not in an authoritarian way. The problem is that this *Gemeinschaft*, organic as it was, existed only at the external level due to its economic foundations. It had not yet become a "unity of values." But Staudinger believed that it would become a new cultural "constellation," in which the realization of an individual's life in the *Gemeinschaft* offered an alternative to the "personality" of the mechanistic bourgeois world.[39]

Weber agreed with Staudinger that the individual was more subordinated to the whole in working-class communities, but he challenged Staudinger's contention that individualism in its own right was, ultimately, *gesellschaftlich*. Granted, mechanistic individualism had become separated from any organic whole, but in reacting to this situation, thinkers like Staudinger have overemphasized *Gemeinschaft* and connectedness. Rather one should view both the individual and the community as creative forces operating in conjunction with one another. It is always the individual who creates, and it is always the generality (*Allgemeinheit*) that produces in him. The general manifests itself in the individual, but can never become fruitful if the individual is cut off from it. Weber denied that one can arrive at a simple formula for the individual-*Gemeinschaft* relationship. The internal and external forces acting upon the individual vary as do his or her relationship to the totality.[40]

> It is only the different expressions of the all-one force, whether it is created indirectly or directly from the generality, purely from the stuff of lived experience of the whole or from parts that have passed through the intimate spheres [of individuals] and been transformed by them.[41]

Weber believed that this variance could be partly explained in terms of class structure. He argued that the lower down in the class structure one went, where the "reality" of the "dense" neighborhoods created a "solidarity" in the masses, the more directly one participated in the totality of will and feelings of the *Gemeinschaft*. Here one sees Weber in basic agreement with Staudinger that there was a direct organic relationship between the individual and community among the working classes. But unlike his student, he denied that this organic relationship is necessarily missing in the middle and upper classes. True, the further one moved up the class ladder, the looser the connections of the individual to the totality were, so that at the uppermost level, that of the spiritual leadership, where the personality was most completely developed, there appeared to be isolation of the individual.[42]

But this individual could develop his personality and still share in the will of the totality at these higher levels. The issue was not one of absence or presence but of directness or indirectness. The task of finding his place in the developmental whole had been left to the individual rather than being presented to him directly by the community.[43] As noted above, Weber believed the generality always expresses itself through the individual; the degree of mediation involved does not change this basic fact. Even though there might have been a wide ideological variety among the spiritual leaders, this represented an expression of the antitheses in reality, of the needs and problematic formulations of the times.

> There is...an ultimate unity vis-à-vis all of these contradictions, in whose contemplation they resolve themselves, before which they appear only as various forms of expression of the all-one force, which stamps itself in a changing fashion on the restrictions and conditioning factors of things.[44]

Weber essentially argued that, ideally, connections between the individual and the community become more monadically organic as one moves up the class ladder, but they need not cease to be organic. As long as these organic connections are maintained, the relationship of the spiritual leaders and the masses remains sound. Without them, there is danger of separation.[45]

The fact that this separation existed was due to the mechanistic and rationalistic approach to the world that many had adopted and not to the condition of differentiation itself. This mechanistic approach could also be found in collectivistic forms as well as individualistic ones.[46] In short, Weber refused to equate the individual-community dualism with the mechanism-organism one as Staudinger had done.

> Our real problem is, then, no longer the opposition of individualism and *Gemeinschaft*, but rather the struggle of the vital view of a *Gemeinschaft*, which is carried in the destiny of the masses as well as in the development of the personality, against that mechanism that rationalism produces (and must produce because it operates only with numbers and their sums), regardless whether rationalism is wrapped individualistically or collectivistically.[47]

In this series of essays on youth and class, one sees the same vitalist, monadic assumptions present in "Religion and Culture": that the centers for cultural renewal have to develop within individuals, and that there will be some sort of cultural harmony among the centers, because they all participate in life through feeling. The old rational forms, those of both the Historical and Austrian schools, were incapable of aiding in the creation of these centers and thereby of solving the cultural crisis. Weber hoped that the new discipline of cultural sociology would help bring clarification to this situation.

The sociology of culture

Weber's writings during the empire on the discipline he termed cultural sociology are brief or sketchy. From the lecture course that he gave twice from 1910 to 1912, only a set of student notes has survived. In late 1912, he presented his new discipline in the opening address before the German Sociological Association. "The Sociological Concept of Culture" was intended as the preliminary statement of a more inclusive work that would build on a series of empirical studies done by his students.[48] In addition, he published a two-page announcement of that series and a short 1915 rebuttal of criticism by Georg Lukács. In these works, one can see the same incorporation of vitalist elements that was present in the essays already discussed. While there were some changes in the discipline in the succeeding years, the general pattern remained the same for the rest of his life.[49]

Weber organized his cultural sociology upon the dualism of civilization and culture. In his lecture course he introduced a third term, the social process (which worked in conjunction with the civilizational process).[50] It was subsumed in the latter in his 1912 address, only to be restored as a separate entity after the war, although it would never achieve the importance of the other two. To establish the parameters of this dualism, Weber turned to the vitalism of Bergson and Driesch. Both had tried to solve a similar problem within the natural sciences: how to recognize the success of positivistic methods in fields such as chemistry and physics and still allow for the organic unity of life. Weber used each for a specific purpose—Bergson to organize the dualism and especially to delineate the civilizational process, Driesch to describe the cultural sphere.[51] He departed from both by transplanting their biological models into the cultural sphere. In defining "life" as cultural, he reduced the biological sphere to the satisfaction of physical necessities and assigned it to the civilizational process.

To distinguish between civilization and culture, Weber used Bergson's distinction between intellect and intuition as a model. He depicted the civilizational process as the intellectual and technical realization of the natural requirements of humanity. A "cosmos of general objective actualities" that can be assimilated but not personally amended, it is essentially utilitarian, instrumental, and characterized by the development of the intellect in order to master the external world. This mastery called for rational analysis in which things are mechanistically related to one another. Within this limited intellectual sphere, a course of progress can be charted, which was why Weber described civilization as a process. While its development does not occur at an even pace in different times and places, as a whole it moves progressively forward.[52] This process is akin to, if not identical with, the rationalization process, which Weber, like his brother Max, saw as most advanced in the Western world.

Again bowing to Bergson, Weber used the term "intellectualization" to criticize those who conceptualized all of the spiritual activity of humanity in terms of the civilizational process. By imposing universality on this activity, theories of history such as those of Hegel, Marx, Spencer, and Comte did not allow for either the individuality of creative activity or the unity of meaning and valuation. These latter concerns make up the cultural sphere, which is organized by an aspiration, or will (*Wollen*), to overcome the analytical division of the world and to grasp the unity that is sacrificed in the process of intellectualization. Weber termed the center of this unity "life feeling." He argued that this center could not be grasped by rationally analytical ("intellectual") methods. Rather, it remained in the hands of feeling, that is, intuition.[53]

At this time, Weber, while certain about what culture is not, seemed less sure of exactly what it entails. He defined it as being present "when life transcends its necessities and utilities to a formation situated above these."[54] Unlike civilization, which was consistently defined as a "process," culture was portrayed in various places as "emanation," "aggregation," "dynamic," "appearance," and even "process." Only after the war would he settle on "movement." In keeping with his other writings of the period, Weber saw culture as monadically organic. He was especially concerned with the relation of the individual personality, particularly the creative personality, to the larger cultural entity. Culture and cultural performance are that which the "will to unity" creates when it is "directed toward the totality of our own internal being united with the totality of everything that faces it from the outside and thus represents the synthesis of personality and world."[55]

Individuals engage in their own personal actions, but their cultural products, especially the creation of works of art and ideas, give concrete expression to the life-feeling of that culture, a supraindividual unity. In a later defense of his concept, he would elaborate further:

> All sociological projection of cultural phenomena must be brought into relation to an ultimate *central point*, which to a certain extent signifies the unified consolidation of the hierarchical order and orientation of the various levels that one projects. In order to find an expression for this central point, I have labeled it the "life-feeling" of a certain era. That is a pure and (intellectually viewed) doubtless an "unclear" heuristic concept, nothing other than the designation of a preintellectual center of lived experience that is present in every era, whose individual features can not be intellectually enumerated, but which is directly intuitively apprehensible in its essence. It is apprehensible with complete precision just as is, on the other side, the essence of the cultural phenomena facing it. The essence of this life-feeling, its quality *with regard to content*, is as little considered by sociology as is the ultimate artistic or metaphysical essence of the corresponding works of art and ideas.[56]

Weber wrote that the relationship between the concrete cultural phenomena created by individuals and this organic center is the subject matter that distinguishes cultural sociology from other disciplines. Its task is also to relate both of these elements to the civilizational process and its objectifications, with which they interact. In so doing, cultural sociology moves beyond sociology proper. The role of the latter is the formulation of social types; the role of cultural sociology is to relate those types to the "central point," the life-feeling of an era. He warned against trying to collapse those two operations into one, which leads to the attempt to explain culture using categories designed for the analysis of the civilizational process.[57] When this happens, people perceive a loss of meaning, an inadequacy of valuation, which characterizes cultural crisis.

Weber's civilization-culture dualism is a variety of the mechanistic-organic one to which many members of his generation (and of the preceding generation) were wedded. Weber attempted to formulate his version in a way that avoided what he saw as the main shortcomings of others. As noted above, the two primary types of this dualism were Hegel's, which was hierarchical, and Tönnies's, which was successional. The Hegelian type subordinated the mechanistic realm (civil society, which includes economic activity) to the organic sphere of spirit (which is embodied by the state). This hierarchical model formed the essential premise of Schmoller's generation of the Historical School and the larger discursive coalition. The problem was that economic life, which the Historical School subordinated to the state, was increasingly seen as the dominant force in the world, signified by the younger generation's terming of the era as "capitalistic" (something Schmoller's generation refuses to do) and its relegation of the state and its bureaucracy to the mechanistic side.

The difficulty with Tönnies's successional *Gemeinschaft-Gesellschaft* dualism, for Weber, was that it forces one into a unilinear teleological position. Even though Tönnies bases his dualism of social interaction on a second psychological one, he basically saw *Gemeinschaft* giving way to *Gesellschaft*. This meant that the sphere of meaning and values would become rational and mechanistic. This model allowed for three scenarios. First, one could accept this succession as inevitable and advocate a new ethic corresponding to it, something most did not attempt. Only Max Weber tried to arrive at a *gesellschaftlich* type of valuation, and without the optimism that normally accompanies this approach.[58] Second, one could adopt the opposite position—a cultural pessimism of decline—which was more popular outside the academy than inside, at least during the empire. Third, one could hope for the return of *Gemeinschaft* in a new form. Tönnies and, as we have seen, Weber's student Staudinger took this stance, believing that the working class could bring about this new form. But this meant overcoming *Gesellschaft*, or somehow subordinating it, which put one back to the hierarchical model.

Alfred Weber understood the problems inherent in the successional model. Therefore, he tried to formulate the dualism in terms that were neither hierarchical nor successional, by making historical change multivalent rather than unilinear. (His essay began with an attack on reductionist, teleological approaches to history.) For Weber, the civilizational process represents the progressive, mechanistic change in areas such as economics, science, and technology. The cultural sphere represents change that is not progressive and allows individual creativity to be organically tied to communal values. With this approach, one can have concurrently both mechanistic progress and organic collective meaning and values with neither sphere subordinated to the other. Of course, Weber believed that these two spheres do interact with one another, but not in a way that is hierarchical or successional.

The early published reception of Weber's cultural sociology was largely negative. There were two significant critiques of "The Sociological Concept of Culture," one by Walter Köhler in Schmoller's *Jahrbuch* and one by Georg Lukács in Max Weber's *Archiv*.[59] Both of these critiques took a conventional position with regard to sociology—that it is more a specialized method than a body of content and that its main task consists of examining objects in their social relationships and not in trying to locate some deeper center that gives meaning to an era. In a sense Weber also took this position with regard to sociology proper, which is why he tried to establish cultural sociology to move beyond the limits of that discipline. Both Köhler and Lukács denied that such a discipline could exist or, if it did, that it could not be called sociology. That cultural sociology does not have a methodology inherently different from sociology proper determines what Lukács saw as its basic question:

> Which new points of view arise if we consider cultural objectifications as social phenomena? Expressed in terms of transcendental logic: what is changed in meaning, content and structure of cultural objectifications if they are invested with the methodologically-sociological form that allows them to appear as social products and thus as objects of sociology? Sociology, like every method and every science, is a form and not a field of objects or content.[60]

Köhler argued that Weber's cultural sociology had abandoned the realm of scientific inquiry by emphasizing feeling over rational methods. He counted that Weber used terms for "feeling" (*fühlen*, *spüren*, *empfinden*) 47 times and the phrases "I think" and "I know" only one time each. If sociology is grounded in feelings, it can no longer be called a science. "Science *thinks*, it relinquishes feeling to other spheres." Köhler made two related points in his attack. First, feeling is always subjective; feelings might be true, but they can never prove their truth. By sacrificing the objective for the subjective, Weber was, in effect, advocating relativism. Second, sociology cannot focus on the unique as Weber advocated; that is the task of history. Weber might have labeled his discipline "sociology" but in fact it was not.[61]

Köhler's concerns were largely reiterated by Lukács in a review of Staudinger's book, which, he wrote, demonstrated the weaknesses inherent in Weber's description of cultural sociology. By using concepts such as "life feeling," "dynamic outgrowth" and "constellation,"

Weber was unable to elucidate a pure cultural sociology and instead surrendered it to a mixture of "intuitive-historical" elements. A definitive sociology of culture is possible only when a set of principles is connected with empirical investigations. Among the questions that such an approach has to include are: which social forms appear as factors that influence cultural objectivations, and how far do these social forms extend into cultural objectifications as formative factors? Weber, by introducing notions like "life feeling," never made these connections. His approach, which was too broad, incomplete, and arbitrarily subjective, simply floated over the facts without making any real organic connection with them. Staudinger's work, in Lukács's mind, demonstrated this failure by placing general principles and empirical research side by side without ever making the connections between them.[62]

In response to Lukács, Weber argued that the latter tried to reduce two separate planes, one abstract and sociological the other concrete and historical, into one, the former.[63] In other words, Lukács (and Köhler) conflated cultural movement with the unilinear and universal civilizational progress.[64] Weber argued that that the multiplicity in the cultural sphere, when viewed from a sociological standpoint, is a "pure historical contingency." It cannot be explained through placement in a causal chain of analysis. Since the historically concrete plane cannot be reduced to the sociologically general, it is necessary to posit a "central point" that provides unity and orientation. This central point for a specific era is what he meant by "life feeling."[65] Weber did not challenge the criticism that such a position is inherently relativistic, which he had already anticipated in "Religion and Culture."[66] He understood that such a charge could be made by anyone such as Lukács and Köhler who did not accept his monadic organic premise.

But how does one go about sorting out the relationship between culture and civilization? How does one relate Weber's rather vague statements about life-feeling to the concrete phenomena of people's lives? In other words, what is the methodology of cultural sociology? Weber had no developed answer at this time. He rejected the notion that the methodology can be developed in advance of specific cultural investigations.[67] Accordingly, he organized a series of specific studies, the first of which was Staudinger's above-discussed book on associations. In announcing this series, Weber designated three areas of cultural social relationships to be examined:

(1) *Cultural organization*, i.e. the construction and the essence of external formations in which cultural movement takes place.

(2) *Cultural interests and cultural productivity of the social strata*, i.e. the differing actually vital spirituality of the different parts of the population. (3) *The life currents*, i.e. the tendencies in the economy, technology, politics, religious organizations, etc. that have directly comprehensible cultural significance.[68]

In 1912, the same year as his address on cultural sociology, Weber applied his new dualism to the issue of Jewish assimilation in Germany. This piece was written as a part of the public discussion that accompanied Werner Sombart's book, *The Jews and Modern Capitalism* (1911). Sombart argued that Judaism, rather than Max Weber's Puritanism, provided the origins of the capitalist spirit, and he identified the Jewish character with both the positive and negative elements of capitalism—great productivity and an economic egoism that was materialistic and calculating. In this book and another one, *The Future of the Jews* (1912), Sombart held that because the Jewish spirit was so different from the German spirit, the Jews were unable to fully assimilate successfully. He urged Jews to maintain their Jewish identity, which he believes would be a greater contribution to Germany than would attempted assimilation. Sombart's books and the lecture series that preceded them were the subject of much public debate by Jewish assimilationists and Zionists. Sombart's audiences were largely Jewish, with assimilationists mostly responding negatively and Zionists, who believed in a separate Jewish national identity, mostly positively. As a result of the attention Sombart's books and lectures drew, Adolf Landsberger organized a volume on Jewish assimilation. Weber was one of those who contributed to this book.[69]

Weber argued that Jewish character, like that of all peoples, can be divided into its civilizational and cultural parts. Peoples with a longer history have longer involvement in the civilizational process and, thus, will be more highly intellectualized. Since the Jews are a much older people than are the European peoples, their level of intellectualization is greater. Sombart was correct, Weber writes, in noting the relationship between this intellectualism and modern capitalism, for both are part of the civilizational process, but he was wrong to attribute this intellectualism to the Jewish religion, which is a part of the cultural sphere. "The Jew is rational, not because he is a Jew, but because he is old."[70]

Weber did not deny that the impact of Judaism on Jews was strong, but he asserted that this impact lay solely in the cultural sphere. It affected their art and their ideas, but not their practical-organizational behavior. He also argued that only this sphere would be significantly affected by assimilation. If Jews maintained their Judaism, then a separate Jewish culture would remain alongside the German culture. (He did not believe this would pose any danger to Germany.) If Judaism eventually disappeared, then so would the Jewish culture, and assimilation would be possible.[71] Here Weber seems to contradict the position he took in "Religion and Culture" that religion need not be the center of a culture. If he stayed with that logic, he could have claimed that Jews and Germans shared a common cultural center other than religion and that this center could unite different religions.[72]

Weber denied Sombart's contention that Jewish intellectualization would have an impact of German culture. Intellectualization is not a peculiar Jewish characteristic; it arises and increases in all those who participate in the civilizational process. With the development of economic life in Germany, intellectualization would spread whether or not there were any Jews. The assimilation of the Jews on the civilizational side would occur automatically, since they would simply have the same universal characteristics as the Germans and any others who participate in that process.

> The tensions that are present [between German Jews and Christians] have always grown out of the civilizational deliberation of the Jews and their external position in life. As they are strongest in those places where the distinctions in intellectualization are the greatest, so they will slowly disappear with the progress of these and the equalization of life position that results.[73]

While Weber's position aligned exactly with neither side in the assimilation debate, his argument clearly favored the assimilationists.

This minor piece, ignored in the critiques of the new discipline and in the assimilation debate, is significant for two reasons. First, it is much more specific in defining the cultural sphere. Vague notions of life-feeling give way to a discussion of a particular culture, even if that culture is not empirically investigated. Second, Weber brought his monadic organic premise directly to bear on a specific policy issue. In this sense, the essay anticipated the direction of many of his writings. The most prominent of these are his discussions of the cultural dimensions of politics before and during World War I. During the last years of the empire, he wrote less and what he did write focused on domestic and foreign policy, although it implicitly followed from the same organizational premises as his cultural essays. It is to those political writings that I now turn.

5
The Cultural Theory of Politics

We have seen that Alfred Weber's career began under the auspices of Gustav Schmoller within the framework of the Historical School of National Economy, the Social Policy Association, and the larger discursive coalition. The focus of this field was the imperial state, which was seen as the embodiment of the organic "spirit" of the nation and hierarchically situated above the materially and mechanistically divided civil society. Without being anchored by the state, the public sphere, where national dialogue and the establishment of values occurred, would become as fragmented as the larger civil society. At the center of the anchored public sphere was the cultivated public, and at its center was the academic elite, which was, at the same time, a part of the state bureaucracy. The Historical School, foremost among the leaders of the academic side of the discursive coalition, dedicated itself to Social Policy—the merger of social science and state intervention, to forestall the fragmentation of the larger public sphere. This strategy focused on the "social question," the integration of the working classes to their proper place in the discursive hierarchy by ameliorating the material sources of their alienation. Weber's studies of the domestic system, like his brother Max's studies of East Elbian agricultural labor, were part of that strategy.

In the first two decades of the twentieth century, both Weber brothers took roughly the same position regarding the imperial political establishment, especially the authoritarian bureaucratic state. Both advocated an increased parliamentary role. However, their positions with regard to the relationship between the state, academia, and the larger public sphere differed. This divergence was less evident before World War I, when they were focused on attacking the presuppositions of the older generation. When it came to establishing their own preferences as the imperial state staggered in the final years of the war, their differences would become more evident. These will be discussed in Chapter 8.

This chapter examines Alfred Weber's political writings during the empire, the most formative period of his career. He organized his political ideas, both implicitly and explicitly, upon the civilization-culture dualism of his cultural writings. For him, the bridging of these two realms formed the central task of politics, which would be aided by cultural sociology. Thus, his new cultural theory provided the premise for his political challenge to Schmoller's version of the state and Social Policy. During his entire tenure in Heidelberg, Weber consistently viewed the bureaucratic state as a civilizational impediment to the establishment of organic connections between individual citizens and the larger cultural values of the German nation. He advocated that the leaders in that endeavor should be the cultural elites.

His cultural views also shaped his foreign policy, especially during World War I, when he viewed Germany as the potential spiritual leader of a group of nations that were politically autonomous but culturally linked. In this monadic scenario, which mirrored his domestic views, he followed Friedrich Naumann's strategy of *Mitteleuropa* (Central Europe). Whether his policy focused on issues within Germany or on foreign relations, the same structural pattern held.

Challenging the bureaucratic state

In 1907, three months before Weber received his call to Heidelberg, Gustav Schmoller, as a representative of the imperial discursive coalition, published a piece in a Viennese newspaper in which he defended the specific personalities and policies of the German government and the "constitutional" (monarchical, bureaucratic) system in general. He argued that the bureaucracy provided an element of stability in times that were threatened with crises. Government officials were cultivated and objective, acting in the interest of the entire nation, not just that of a single class. They were

> the class of society that prepares itself vocationally for public business, that has learned to lead the state administration by rising up the ladder of offices, that does not stand within the machinery of economic life, that does not have making money as its main goal in life, that recruits itself from all classes of society from the highest to the lowest.[1]

The alternative, parliamentary government with ministries formed by the ruling political parties, was inherently unstable because those parties answered only to specific class and religious interests. "They are increasingly class and religious organizations that combat the state and the public weal to gain advantage for their class or religion." Because of this "class egoism," any coalition of parties was only temporary and would dissolve if the party interests so dictated. None of these parties looked to the nation as a whole.[2]

Two weeks later, Weber responded in the same newspaper to Schmoller. He wrote that Schmoller's article should be read as a "defense" of the existing system and, thus, as a negative answer to the question: "should a transition from the current system of government by officials to one of party ministries take place, ... should one replace the present so-called constitutional regime with a parliamentary government?"[3] Weber proceeded to answer this question in terms of the "cultivation of political will," which "is consummated in the parliamentary system by parliament and in the constitutional system by the authoritarian power, in the first case it is consummated by the people (*Volk*), but not in the latter case."[4] By separating the will of the people from the authoritarian state, Weber challenged the formulation of the discursive coalition that the state was the highest expression of the will of the people.

Weber assigned to Otto von Bismarck prime responsibility for the shattering of parliamentary forces in Germany after 1848. The system that was constructed in these years reflected the will of only two men, Bismarck and the Emperor. Those who were grouped around these two—bureaucrats, nobles, military men—had "no legitimate form to create a cultivation of political will." They were simply functionaries. However, the parliamentary forces could not be permanently suppressed, hence the current struggle of the government with forces in the *Reichstag*.[5] In other words, the constitutional state was designed to suppress the will of the people. When that will expressed itself in spite of the system, the ruling elite tried to make the system even more authoritarian. Such a strategy, Weber implied, went against the times.[6]

Weber did not deny that parties represented class interests, but argued that a coalition of class parties did not make a class government. Parties, on entering a government, had to accept governmental responsibility and compromise with their coalition partners, allowing them to act in the interest of the larger nation.[7]

In 1909, at a meeting of the Social Policy Association, and in *Die neue Rundschau* the following year, Weber expanded these negative comments about the bureaucratic state, tying it to the "monstrous process of rationalization."[8] In these two forums, the bureaucratic state was identified with three forces to which the discursive coalition had considered it antithetic: technical specialization, the capitalist economy, and political parties.

Weber claimed that the modern official was characterized by the narrowness of a specialist. Officials had become technicians with no "values of feeling." The bureaucracy had the tendency to absorb the natural elements of life into its "chambers, disciplines and sub-disciplines," submitting it to the "poison of schematization," replacing that which was unique and alive with a "gigantic, mathematical something," the workings of which were "inanimate," "fragmented," and "soulless."[9] Schmoller had seen specialization as an important characteristic of officialdom, especially academia; however, he assumed that this specialized activity was always informed by a larger set of organic values that characterized the state as a whole. Weber did not deny that such organic values existed, but denied that they were embodied by the state. For him, the bureaucracy was devoid of any meaning, officials being simply functionaries.[10]

Weber described the bureaucratic state as part of the same rationalistic process as modern capitalism. As an example, he pointed to the transportation system that he had theorized about in *Location of Industries*, where "not fewer than 150,000 bureaucratic functionaries have to participate alongside a half million workers in this apparatus in Germany today."[11] Weber argued that state officials were, in fact, no different than the so-called "private officials" who worked in banks and other large capitalist concerns.

> There is no doubt in my mind that all of these public functionaries will at one time fall in with the other, private ones. Those entire strata of our middle classes that have been transformed by bureaucratization and that must continue to grow with every stage of the mechanization and rationalization of our existence, the complex of engineers, technicians, businessmen, etc. who stand within the apparatus and within it today undoubtedly perform the most important and most modern work—they all have already been organized and their organization, which seeks to build the foundation for their new appointment in life, will also at one time be the center of attraction for the organization of state officials.[12]

This is an ironic twist on perceptions of the time. Employees in large concerns called themselves "private officials" in order to share the prestige that state officials enjoyed in Germany.[13] Weber reversed the relationship and undermined the privileged position of state officials by describing them as essentially no different than employees in capitalistic concerns. Both state and private bureaucracies operated according to the principle of the division of labor; both restricted themselves to the "practical-utilitarian" and ignored broader spiritual forces.[14]

> It is merely "historical chance" that *capital* assumed control of this intellectualization as a means of realizing profits and in doing so oversaw the establishment of a new organization. It could just as easily have been the *state* (as has in part actually been the case) that undertook the general rationalization. However, because the latter is there, it today leads to general bureaucratization.[15]

Finally, Weber, like his brother Max, denied that officials were above party interests. It was a mistake to think that the state bureaucracy did not have a distinct social base. It was clearly tied to the ruling classes, membership in the bureaucracy being based on membership in certain parties. As a result, officials were not objective individuals who looked to the nation as a whole, but rather they were promoters of party interests at the expense of the rest of the nation. Weber labeled this tie between officials and party interests a "force for the corruption of conscience." He called on the Social Policy Association to demand a sharp separation of the political party apparatus from the bureaucratic apparatus. [16] Here, one sees Weber taking a somewhat different tack than that of his earlier response to Schmoller. Then, he had accepted the dualism of bureaucracy and political parties but challenged Schmoller's privileging the former. Now, he lumped both along with the process of technical specialization and the capitalist economy on the same side of the larger organic-mechanistic dualism.

Weber also reformulated that cornerstone of the discursive coalition, Social Policy and its "social question," the integration of the working class into the nation. In fact, he wrote, the current policy simply attempted to extend to the working classes the same process of bureaucratization that had already captured the upper and middle classes. "Everything that we have so far called 'the social question' is nothing other than the mirroring of this process [of rationalization] in the content and destiny of the life of the lower strata through their absorption in the mechanism." It took these people out of their previously "free existence" and turned them into mere labor power, thereby throwing them into "that gray, barren, uniform casing (*Gehäuse*)" of bureaucratized existence.[17]

Formulating a new Social Policy

In 1912 and 1913, Weber called for a new Social Policy in three articles, all of which were published in Max's *Archiv* rather than in Schmoller's *Jahrbuch*.[18] In a manner parallel to his addressing of the cultural question, he argued that a new historical situation had begun to break up the old sociopolitical "constellation," which had been in existence for the last 50 years. The problem of the old constellation involved not only the incorporation of the proletariat into the changing economy, but also the struggle between traditional and capitalist economic forms. Both these issues had a common concern, economic egoism, which was seen as the cause of most of the misery. Those forces, both conservative and socialist, that opposed this egoistic capitalism, looked to the state as a countervailing force. Because the state was seen as the embodiment of the principle of a common economy, the need to address the problems of the working class under the old constellation actually strengthened the state.

That view, Weber felt, was no longer valid. The state had proven itself to be the greatest industrialist in the capitalistic system, with the result that the working class was even more enslaved than before. The state did not create a true community (*Gemeinschaft*), but rather the compulsory congregation of different classes through its authority. Instead of the freedom of the masses, the goal of the current policy was simply that of "reason of state." Weber called for a new set of goals that would allow the worker to determine his or her own "vocational (*Berufs*) destiny."[19] There were two sides to this freedom, the material and the spiritual.

Any effective reform program had to address the fate of individual workers, rather than treat them as identical cogs in the machine. Weber noted that while the system attempted to bureaucratize the workers by restricting their independence, they did not enjoy the same "vocational curve" as the official. Rather than improving their lot after the age of 40, many workers actually saw a decline in their standard of living. While the official received increased security, the worker experienced increased insecurity. So while workers were subjected to the regimentation of the official, they did not receive corresponding material rewards. Among the reforms that Weber suggested was "free private insurance" to liberate workers from the "configurations of coercion" of the state.[20]

The improvement of material well-being, while important, was not the central problem for the integration of a new class.

This central problem is the manner of its spiritual incorporation, the rescue of its personality and freedom from the mechanism by the maintenance of the discipline necessary for their functioning. And so, from the problem of the emancipation of the worker, the employee and the official, the present social question arises as the question of the rescue of the personality from absorption in the apparatus.[21]

The key to this reform program was the development of the workers' ability to determine their own vocational destiny. That meant better education in their vocation, the addition of variety to the vocation in place of simple monotonous routine, and even provision for the opportunity to change jobs.[22] In short, reform had to keep workers from becoming as spiritually narrow as officials.

Weber called for a new kind of organization that would set the personality free in life "through the support of all democratic forms of labor rights and through the fostering of all possible internal democratic arrangements of the various organizations, even workers' organizations themselves, against their internal bureaucratic and autocratic tendencies."[23] One example of Weber's program was his opposition to the "yellow unions" set up by the employers themselves to eliminate the workers' right to strike. He described these unions, found mostly in huge industrial firms, as a symbol of a general spiritual condition, the bureaucratization of labor. Being organized according to the plant rather than the vocation, they isolated workers from others in the same vocation in other plants, preventing the development of any "vocational solidarity." The result was worker docility; workers were turned into pseudo-officials with little independence. Instead, Weber advocated continued support of the socialist union movement. However, he emphasized that these unions had to work to develop "a free spirit unaffiliated with parties" in their workers.[24]

Even in attacking yellow unions, Weber attacked the bureaucratic spirit and, with it, the discursive coalition. One factor that stood in the way of a new Social Policy in Germany was the worship of officials, accompanied by a "metaphysic of bureaucracy," not found in France or in the Anglo-Saxon countries. It was better to have the corrupt officials of America, whose limitations were recognized and thus who left room for the development of the free personality, than it was having to bow and scrape to a ruling clique. He accused the older generation in the Association of promoting just this negative kind of acquiescence.[25]

Weber believed that moderns, especially Germans, could never be totally free of bureaucratic organization, but that they could do better in moderating its impact than they had. This meant attacking the tendency to treat the state "apparatus" as a "mystical something." "It will be an important consequence if the metaphysical veil of the official falls away because of a newly created valuation of the vital force in people and the atmosphere of life that this valuation produces."[26] The key was the concept of vocation, or calling (*Beruf*), which was akin to the ideal of cultivation (*Bildung*).[27] The center of both was the development of the personality. Vocation, which Weber, following his brother, attributed to the inner-worldly asceticism of the Puritan, had to be retained but with "a new basis, a new content and new limits."

> Such a meaning will probably be there anew, and apparently always be there: the confirmation of the person in life through activity in life. But today we will say: confirmation of the person, not through a faith of some sort or an object—the person as the ultimate, only ground, the only force in life, as the source of all great things for which we can sacrifice ourselves. For we want a life filled with such *great* things and nothing more. The goal of the development of the person within the vocation, however, will at the same time provide for us the limits of our dedication to the vocation.[28]

How could a new foundation be established—one that would allow the development of the person? How could the person be freed from the "dead mechanism" of the apparatus? Clearly Weber believed that this could not be thought out through the old formulas that characterized the discursive coalition. Those formulas were seen as—at best—ineffective, but, more importantly, as part of the problem. A diagnosis of the situation, which Weber, along with many of his day, defined as a cultural crisis, became the central task of his new discipline, cultural sociology. Not that the discipline could provide the answers. Rather, its role would be to map the discourse in which the answers were sought. A discussion of how that would be realized, and the differences with Max that it entailed, was postponed with Germany's entrance into the war. Here, he offered only a diagnosis and not a prescription.

Weber's diagnosis was defined using the vitalist concepts of his theory of culture. Vital life forces existed in the individual and developed along with his or her personality, which was organically in tune with the larger cultural whole. The bureaucratic state imposed artificial forms on the life forces, which hindered their development. These monadic organic premises would also shape Weber's views on international relations.

World War I and the ideal of *"Mitteleuropa"*

With the outbreak of the war in 1914 and the perceived emergence of a "national community" (*Volksgemeinschaft*), Weber's emphasis shifted from domestic policy to foreign policy. A significant portion of his published wartime letters to Else Jaffé deal with understanding Germany's place vis-à-vis its opponents. He argued that only by defining Germany's "world tasks" could the country decide with which of its opponents to settle its differences and with which to continue its fight.[29] The criteria for this decision were cultural. Which of Germany's opponents posed the greatest danger in a period of cultural upheaval? This collection of letters was largely written and published in 1915, when the course of the war, with its emphasis on the western front, had already been established and the deterioration of the German position domestically had not begun. There appears to be little correlation between the fortunes of the German war effort and Weber's positions regarding which of the Entente was the main threat, culturally.

Weber believed the war represented a watershed, a period of tremendous change that could not be comprehended by the existing cultural concepts. Six months into the war, he wrote:

> Our old conceptual universe, words such as democracy and aristocracy and the like, are absolutely useless vis-à-vis our present realities and can no longer remedy the contradictions at all. The war has placed before our eyes the new facts, the actual realities and forces that are present, so that we can now articulate them in a new world of words and concepts.[30]

How would the different countries respond to this cultural challenge?

He admitted that the Germans were barbaric when measured by the "standards of civilization," because their nature was so strong and they were able to see and do what was essential—they would be the destroyers of old conventions and forms, so that something deeper could be reached. "There is present a great and original connectedness and unity with the eternal and direct source, which ascends in all of us Germans." They were called upon to usher in the new world.[31]

What we can offer to the world is no longer forms of political organization that must be completely adjusted to our unique international situation. What we want to offer the world is also not capitalistic or pseudo-capitalistic domination, but rather—here the philosophizing chancellor has really coined a good word—an "imperialism of spirit!"— or better, a "gospel of spirit!"[32]

Opposed to this spirit stood Britain and Russia. (Like many at the time, Weber was less concerned with France.) Weber judged these two powers on the basis of whether they presented a threat to Germany that was internal, that is, cultural, or merely external. Although Weber did not formulate these judgments explicitly in terms of his culture-civilization dichotomy, the latter was clearly in the background of his statements. At first, he believed that because the Russian world was spiritually so different from the German one, it presented no internal, cultural dangers to Germany. In the future there would be a significant rivalry between the Russian world and the European one, but, for the present, Russia offered only an external danger to the Germans. Therefore, he advocated that some kind of settlement should be made with Russia in order to concentrate resources on combating Britain. The British-American world, on the other hand, was a variety of the European essence and competed directly with the German spirit within that cultural entity. Hence, "Anglification" presented an immediate danger to the European world, over which Germany had to gain spiritual mastery before it could lead a cultural struggle of that world against Russia.[33]

Within three months (by May, 1915), Weber's view of the situation had changed. Now "the colossal, fertile, sleeping Russia" became the primary opponent. In making this switch, he became more consistent in his use of the unexpressed culture-civilization dualism. Britain, whose values were essentially civilizational, was really a threat to Germany only if that external civilizational realm could eclipse the cultural realm. But it offered no competition within the cultural realm. Russia, whose essence was primarily cultural, was the main threat to Europe's internal cultural being.

Weber believed that the British spiritual danger would not extend beyond the war. The British challenge was essentially materialistic, one of imperialism and commercial interests, a "cold egoism." Britain would remain something great as long as capitalism thrived; however, its continued existence would be no threat to deeper German spiritual values. "Internally we can learn nothing from the English and can have no effect upon them." In fact, as the end of the age of materialism loomed near, it was clear that the British and their French allies had exhausted their essence. Accordingly, Weber recommended a strategy of containment of British sea power, but not its annihilation.[34] Britain's lessened prestige would weaken its negative spiritual influence in the world and hasten the end of the era of materialism. Weber suggested that emphasizing competition with Britain in the material realm of power would favor the power politics (*Machtpolitik*) of nationalists like Treitschke, who believed that might made right. This approach would put the primary emphasis on the state and thus would divert Germany from its larger cultural mission. While Germany had to maintain its world power to preserve its spirit and law, these political motives were properly always subordinated to an Idealism of "the Word."[35]

Why did Weber make an about face regarding Russia? He attributed it to an April article in the *Neue Rundschau* that warned of Russian "national imperialism" (*Volksimperialismus*).[36] Regardless of whether this article was a deciding factor in Weber's shift, it brought him into alignment with the policy of *Mitteleuropa* (Central Europe) that reached its height in popularity that same year with the appearance of Friedrich Naumann's book by that title.

Naumann's *Mitteleuropa* was the most important work published during the war, with sales of more than 100,000 copies in the first six months. His earlier unsuccessful efforts to establish a liberal political party were supported by the Weber brothers, because he had an inclusionary social philosophy, accepting the working classes as an integral part of German society.[37] This same inclusionism continued in his foreign policy. In Henry Cord Meyer's words:

> Naumann had perceived a fundamental dilemma of our century: with increasing technological specialization and with opportunities for groups and interests to oppose each other, how was the essential sense of unity of a nation to be preserved? How to protect the autonomy and individuality of the human being, and of human values, in the face of the inherently totalitarian challenge of the Machine Age? In his wartime *Mitteleuropa* Naumann was to raise these basic issues again.[38]

Here one can see here a cultural position very similar to that of Alfred Weber, including the need to rethink the current situation. Naumann sounded like Weber when he wrote: "This is the occasion where we ourselves must learn. Wartime is an opportunity for all of us to discover new modes of thought. I speak not dogmatically; I am searching for the future."³⁹ Both men wanted to reconcile individual autonomy with organic unity, in foreign policy as well as domestic policy. In Naumann's *Mitteleuropa* policy, Austrian Germans and Slavs, excluded by the German establishment, were to be integrated into a unity while preserving their autonomy. The policy was antiannexationist, opposing the extension of a greater German state. Naumann wanted a federation of states (*Staatenbund*) rather than a federated state (*Bundesstaat*).⁴⁰ He wrote:

> Let us once again think about the *principal concept of Mitteleuropa*. At its core *Mitteleuropa* will be German and will obviously use the universal and mediating German language. However, from the very outset it must demonstrate compromise and flexibility with regard to participating neighboring languages, because only thus can the great harmony develop that is necessary for a great state besieged and surrounded from all sides.... Here the new youth that grows up after the war shall do better than we elders, so that a type of middle European person, who assimilates of all elements and forces of cultivation, is perfected, the bearer of a strongly diverse and content-rich culture that grows up around Germandom.⁴¹

Naumann's policy competed with two others aimed at different geographical regions, *Weltpolitik* (World Policy) and *Osteuropa* (Eastern Europe). These two policies had themselves been in competition since the 1890s. *Weltpolitik*, which had emphasized the competition with Britain and France for overseas colonies, dominated foreign policy until the eve of the war. As it became less feasible with a series of diplomatic defeats, imperialists and the German military turned their attention toward the continental lands east of Germany. The annexationist promoters of *Osteuropa* wanted to extend German hegemony eastward, especially into Poland. ⁴²

After 1915, the *Mitteleuropa* policy was eclipsed by the *Osteuropa* policy, whose supporters included the chauvinistically nationalist Pan-German League and the wartime Fatherland Party. *Osteuropa* promoters had many more resources, most importantly the government, which had become a military dictatorship of General Erich Ludendorff. The People's League for Freedom and Fatherland, formed by supporters of *Mitteleuropa* in cooperation with many trade unions to challenge the dictatorship and the Fatherland Party, advocated democratization in Germany and national self-determination in

Central and Eastern Europe. Despite a membership of about four million, it did not receive enough support to be effective.[43]

Alfred Weber was an advocate of *Mitteleuropa*, although early in the war his version was more imperialistic than Naumann's. He rejected direct annexation of lands to the East, instead promoting a collection of satellite states with internal autonomy but subordinate to Germany in economic, military, and diplomatic policy.[44]

Weber's foreign policy always had a cultural foundation. In his 1915 book of letters, where his position was softer than his private opinions, Weber wrote that the policy toward the East had to be one not of empire but of cultural hegemony. Only when Germany understood the cultural nature of its world mission, could the struggle with Russia be truly engaged. Unlike Britain, which simply sought to realize business and power interests, Russian imperialism was accompanied by the promise of world salvation. It could only be combated by another Idea, not solely on the level of reality.[45]

Weber argued that the Central European "body" (*Körper*)—a term introduced in *Theory of Locations* and which would be very prominent in his major essay on cultural sociology after the war (1920)—offered Germany the best opportunity to develop itself spiritually and psychically. However, the body "must be an alliance—not a sphere of domination. The fundamentals of imperialism are not applicable to our world membership."[46] Central European nations had be brought within Germany's cultural sphere but not subjected to Germany's political might. In this spirit he also became a founding member of the aforementioned People's League for Freedom and Fatherland.[47] Weber's advocacy of culture over politics paralleled his prewar writings on domestic policy, which challenged the imperial state in favor of cultural politics.

In 1917, in two memoranda and a speech, Weber switched back to his original position about whom Germany should be fighting.[48] The combination of the February Revolution in Russia and American entry into the war, convinced him that the West represented the prime military threat. Although he remained convinced that Russia would continue to be the chief cultural threat to Germany, he lobbied for a "continental understanding," including a reconciliation with Russia, which included a renunciation of any imperialistic designs on Russian territory.[49] In addition to concentrating its military might against the West, Germany would gain two other advantages from the proposed understanding with Russia. First, it could hinder the emergence of a "consolidated imperialistic Russia." In other words, it could guide the foreign policy of the new Russian parliamentary government away from that of the old imperial regime. And second, it could establish stronger, nonthreatening ties to the smaller peoples of *Mitteleuropa*.[50]

The cultural underpinning of this policy was clearly revealed in a lecture of late 1917, which was published the next year. Weber distinguished between the Anglo-American principle of national self-determination, as articulated by Woodrow Wilson, and the German version, which he was presenting. Both principles were parts of the world views over which the war was being fought.[51] Wilson's principle was purely civilizational (although Weber did not use this term). It was abstract, rational, and empty of reality, seeking to impose an abstract uniformity on the peoples of Europe through formal law that ignored historical and geographical realities. Because of the heterogeneity of peoples, a simple abstract formula of self-determination could not be applied without damaging these peoples.[52] The German version of self-determination was based on this historical heterogeneity. Weber wrote:

> In contrast to this uniform life form we place the manifoldness and variegation of our external life, the individualization of our inner life, the gradation of the manner in which we are coined as persons. We contrast the configuration of individuality to the schematic form of unity.—Well, then. A psychic cultural whole, a people also has its individuality and can develop this individuality in democratic times; it can evolve its essence only in accordance with its own law if it "determines itself" in both form and content. No law from above can help it do this, for the psychological forces that are operative in it work in such times not on order from above but only through their own inner impulse, thus "self-determined." According to our view, self-determination is accordingly in democratic times the form, and certainly the only form, of the development of the individuality of peoples. We, as representatives of the evolution of individuality, should hoist its flag.[53]

Weber presented three types of national self-determination. First, there were great powers that, because of their size, geographical location, and economic resources, could change the balance of a region. Germany was an example of this type. Second, there were "national splinters," peoples who are unable to develop a viable separate identity and so they attach themselves to a larger power in order to share in its position (and its identity). An example of this type was Alsace-Lorraine, which was identified with Germany. Third, and in between the other two, were small peoples, who could not change the balance by themselves but could maintain an autonomous identity. Poland was an example of this type.[54]

Britain, Weber argued, did not recognize this heterogeneity and acted like a large industrial firm trying to establish a cartel.[55] In contrast to this attempt to impose uniformity, *Mitteleuropa* would have a different basis in which Germany as the major power would play the major role, but always respectful of the national identities of its neighbors. It would provide the center of a new Europe based on these principles. It would become a spiritual leader. "And accordingly it is, therefore, good and internally substantiated when henceforth we, and with us this entire new world political group, raise the right of the self-determination of peoples as the flag of our lead ships."[56]

Weber cautioned, however, that while self-determination originated from within, it was also conditioned from without. While the one-sided emphasis on economic factors, which had characterized the late nineteenth century, was incorrect, it was also wrong to ignore these factors.[57] Here, one sees the interaction of the two spheres of civilization and culture. Weber argued that the German principle of self-determination was superior because it was historical and recognized the cultural development of peoples. But this "psychic cultural whole" continually interacted with the civilizational process. Economic conditions, if not primary, certainly did help to shape national identity and would be very instrumental in the creation of a new Europe. This represents the same monadic organic pattern that characterized Weber's cultural sociology and domestic politics, in which the individuality of the peoples of *Mitteleuropa* was preserved and even encouraged, with the assurance that this would promote their organic unity. This unity was aided by civilizational factors, the economic ties between these lands. And it was assumed that Germany would provide the necessary spiritual leadership.

By 1917, Germany had clearly become a military dictatorship; there was no end to the war in sight; and privation on the domestic front, particularly in the cities, continued. These factors led to increased civil unrest and an attempt of the *Reichstag* to assert itself. These circumstances brought the Weber brothers back to issues of domestic politics. Both began to develop ideas on what a postauthoritarian

Germany would look like. The brothers' ideas on these matters bore some resemblance to one another, although Max's were more detailed than Alfred's. But there were also some very significant underlying differences, despite the surface similarities. These differences would become more obvious as Alfred's views evolved in the new republic that followed the defeat in war.

Part II
Alfred Weber in the Weimar Republic

6
The Weimar Republic and the End of the Discursive Coalition

As the war continued into its final years, the discursive coalition's formulation of the "national community" proved more and more untenable. Class divisions reemerged with the hardships of the war, and the image of the national community became more populist and relied less on the institution of the state. "Local incidents between consumers and retailers [over food rationing] had escalated into a broad confrontation between citizens and the state. It was not long before the calls for 'bread' were accompanied by the demands for 'peace.'"[1] As the imperial government faded into the background, academics disagreed about how to address the disintegration of the discursive coalition. The moderates sought ways to reconcile the discordant social forces with their own organic cultural aspirations. The annexationist radicals sought ways to repress those same forces and rallied behind the dictatorship of General Ludendorff.

In a series of events in 1917, parts of the government began to align themselves with the populist movement for change. In February, the forces of revolution resulted in a liberal parliamentary regime in Russia. In April, Germany's policy of unrestricted submarine warfare brought the United States into the war. In July, the chancellor, Bethmann-Hollweg, who had been urging electoral reform, was forced to resign. That same month, a "peace resolution" was introduced in the *Reichstag* by the Social Democrats, the Catholic Center Party, and the left-liberal Progressives. It called for immediate peace negotiations based on no reparations and no annexations. This came to nothing as the military strengthened its hold on the country, supported by a large propaganda organization, the Fatherland Party. Nevertheless, civil unrest continued, resulting in a wave of strikes in 1918. In the spring of 1918, Ludendorff gambled on an all-out offensive on the western front. Its failure and the unrest eventually resulted in the overthrow of the regime by democratic forces in November.

The November revolution, and the republican constitution ratified less than a year later, ended the Weber-Schmoller debate in Weber's favor: Germany became a parliamentary nation with power centered in the *Reichstag*. An elected president was given special emergency powers, but any exercise of them had to be countersigned by the chancellor, whose retention was dependent on a Reichstag majority coalition. The original "Weimar coalition," composed of the oppositional parties of the empire (the SPD, the Center, and the liberals[2]), continued in one form or another until its break-up during the first year of the Great Depression. This amounted to the same coalition of class parties that Weber had advocated in his 1909 critique of Schmoller's "constitutionalism," and his own Democratic Party was an instrumental part of it.

With this transformation, half of the old discursive coalition, the authoritarian state, disappeared. The question of the significance of the other half, the spiritual leaders (especially academics), and its relationship to the new political structure became a central issue for intellectuals like Alfred Weber. The question was made more pressing by the great uncertainty that characterized the republic, especially in its early and later years.

The republic has been described as the paradigmatic site of the "crisis of classical modernity," a crisis characterized primarily by the increased fragmentation of both the political and cultural spheres. Most of the characteristics ascribed to classical modernity—organized capitalism, urbanization, the growth of modern political parties, the welfare state, and modernist culture—originated in the empire, but all these tendencies accelerated during the Weimar period.[3] To the difficulties associated with these changes was added the physical and psychological devastation wrought by the war.

The economy contributed greatly to the political and cultural dislocation in the republic. The financing of the war spawned inflation that, by mid-1922, had been allowed to spiral into hyperinflation, which contributed significantly to insecurity among intellectuals. If the revolution had destroyed the old discursive coalition by removing the imperial political leadership, inflation severely challenged the cultural premises that legitimized it. The inflation, writes Bernd Widdig, represented a radical cultural rupture.[4] In the empire, money was something that concerned the cultivated bourgeoisie but that was not absolutely central to their lives, as long as they were able to maintain a life-style in accord with their status. In Heidelberg, for example, having the wherewithal to entertain was important to one's role within the "eternal conversation" and hence to one's cultural capital. If this criterion was met, one did not fret constantly about money. During the inflation, economic security disappeared, and people who were unused to fixating on money suddenly had to. Everyone knew the dollar-mark ratio. Money not only dominated the economic relationships of civil society, but also extended into other spheres. Thrift, hard work, the morality of the middle classes appeared under attack as there no longer seemed to be a discernible moral order. The sphere of human life that the defenders of the discursive coalition saw held in check by the state and culture now dominated in a carnivalesque chaos. Everything solid melted into money.

In undermining the values of the cultivated public, the inflation diminished the prestige of that public's elite. It significantly narrowed the gap between the latter's less elastic incomes and those of the working classes, thus lessening the elite's ability to demonstrate their status. Loss of economic capital undermined cultural capital. At the time, the most discussed treatment of the plight of these elites was Alfred Weber's *The Distress of the Spiritual Workers* (which will be examined in chapter 8).

Deprived of the legitimization of the imperial discursive coalition, academics faced increased competition from intellectual elements outside of the university who were able to appeal not only to a larger public but also to the cultivated bourgeoisie. Modernist elements of the new, experimental "Weimar Culture" embraced the fragmentation of the old discourses and institutions and sought to build upon it,[5] while reactionaries and fascistic types railed against it. Both groups attracted followers among the cultivated public, to the chagrin of mainstream academics. The latter attempted to uphold the positions articulated during the war and refused to relinquish the hope for a unified sphere of meaning mediated by academic practices. With no consensus on how that would be accomplished, professors were confronted with a larger culture that called the strategic importance of the university into question. Their use of new terms such as "massification of literature" and "culture business" lamented the new mass culture and the media that fostered it. And the political resources used under the empire to repair deficits in coherence and respect were not to be had on the old terms, if at all.

The attack on the traditional academic establishment came not only from without but also from within the institution itself. The new socialist cultural minister of Prussia, Konrad Haenisch, and his liberal undersecretary for higher education, Carl H. Becker, tried to establish a new pedagogical mission to provide civic training for the democratic citizen. Becker proposed a new structure for the university that would give a larger role to students and untenured faculty. He also wanted to modify the curriculum in order to allow for a greater engagement with the sociopolitical environment. This included a modern revitalization of the synthetic ideal of classical cultivation by giving the marginal discipline of sociology a more central place in the university.[6] "The struggle over sociology is really the struggle over the new concept of academic science," wrote Becker.[7] These measures were proposed for Prussia, but the impact was felt throughout Germany. One could argue that, in essence, Becker was attempting to spread the "Spirit of Heidelberg" to Prussia, despite the fact that it was actually declining in Heidelberg at this time.[8]

Becker's proposals were generally unpopular among academics, many of whom saw them as an attack on the autonomy of the universities by the Social Democrats. Historian Georg von Below argued that sociology was not fit to be a synthesizing discipline. Using the language of the old discursive tradition, he claimed that the kind of synthesis Becker advocated was that of "dilettantes" rather than true scholars. The university had enjoyed more autonomy under the empire than it could under democratic forms of government and Becker's reforms amounted to forcing "modern Russian contrivances upon us." At the end of the republic, Weber's Heidelberg colleague, Ernst Robert Curtius, echoed Below's sentiments, denouncing "sociologism" as nihilistic when applied to culture and thus undermining the ability of traditional humanism to resist the growing irrationalism in Germany.[9] As a result of this kind of opposition, Becker and Haenisch were unable to successfully implement their reforms in any comprehensive way.

The Becker controversy overlapped with another controversy at that time over Max Weber's "Science as a Vocation." In both cases, academic values viewed by traditionalists as organic, synthetic, and representative of the German spirit appeared under attack by forces associated with the mechanistic, pluralistic, and materialist civil society, thus reducing the university's resistance to "destructive" modernist forces outside of academia. Alfred Weber did not join either of these two debates directly. However, both were very relevant to his hopes for a synthetic cultural sociology.

State apparatus, elite formation, and the public now operated so as to foster new divisions in the university rather than to create a surface unity, as they had under the empire. The cohesiveness of elites was greatly reduced by the elevation of leaders from beyond the boundaries of the cultivated classes and the creation of alternative institutionalized channels for their advancement. The academic's role of "cultural citizen" depicted by Kruse assumed the presence of a certain kind of public. If, as Schuster noted, that public had been challenged during the late empire, in the early republic it seemed threatened with total dissolution. The theologian Ernst Troeltsch, who, in 1914, had moved from Heidelberg to Berlin, wrote to an American audience:

> We have not yet come to the formation of a normal public life.... In such a state of perplexity and need, there is, generally speaking, no Public Opinion left—only the hope that, somehow and some time, the world will be otherwise.... [T]he best heads in Germany are thinking day and night whether some initiative is not possible here. The educated, intellectual and high official classes [who had composed the discursive coalition] at the present moment have been, unfortunately, almost entirely set aside.... There is anxious questioning whether a solution *can* be found—and whether it can be found in time.[10]

Troeltsch, who reflected moderate academic sentiment, believed the war had destroyed the old values that had been taken for granted, that is, the assumptions that had centered the old habitus. He charged the historical profession with articulating historical values in order to arrive at a new synthesis necessary for modern decision-making. The significance of general world-views and the product of cultivation had to be reconsidered and firmed up. This enormous task was necessitated by the "crisis of historicism."[11] He was not alone in his perception of an intellectual crisis. The term "crisis," popular before the war, became even more prevalent during this period. Writings appeared on the "crisis" of culture, theology, pedagogy, spirit, science, mathematics, physics, mechanics, and causality. "The crisis existed, if only by virtue of the fact that almost every educated German believed in its reality."[12]

The general sense of crisis felt by intellectuals extended to the field of "state law," which included the new constitution.[13] A body of constitutional theory, which examined the relationship of the legal system to the power structure of the state, had been firmly established in the decades before the war. With the demise of the empire, those issues had to be addressed in the new context of social, economic,

104   *Alfred Weber and the Crisis of Culture*

and political turmoil. The viability of the new republican constitution and the parliamentary democracy it established was subject to much debate in the following years, what Moritz Julius Bonn, in a lecture series cited by Alfred Weber, termed the "crisis of European democracy."[14] Bonn recognized that class diversity resulted in social pluralism and he urged abandoning the idea of complete uniformity, although the presence of sound majorities was absolutely necessary for parliament to function. He worried that this was made difficult by economic dislocation, international insecurity, and the constitutional provision for proportional representation. He warned that it was not only the form of government that was important but also the spirit.[15]

Although there were varying views of law and its relation to the state during the empire, the dominant ones fit within the schema of the discursive coalition. The two whose influence continued in different forms with the demise of that coalition were the organic and positivist approaches, whose most influential proponents were Otto von Gierke and Paul Laband respectively. Gierke, following earlier Romanticism and Hegelian Idealism, combined a monistic organic view of the associations (*Genossenschaften*), on which Weber's student Staudinger focused, with the recognition that the state and national law stood above them. He viewed the state as the embodiment of the national spirit (*Volksgeist*). Laband also assumed an organic relationship between the state and the nation. The laws expressed the will of the state, which, in turn, expressed the ethical will of the national organism conceived as a personality. In a defense of the *Rechtsstaat*, he contended that the state made the law, including the constitution, and was bound by those laws. Questions of legal validity stood apart from politics and could be formulated only with reference to the law itself. The authoritarian state of the empire stood above the institutions of civil society, deriving its legitimacy either as an endlessly developing natural-organic person or as the creator of a body of law.

As civil society received its due in the republic, the organic and positivist approaches took on a revised form. Hugo Preuss, who was Bonn's predecessor at the commercial university in Berlin and who authored the first draft of the new constitution, used both approaches. Preuss, who had questioned the authoritarian state early in the war, earning the enmity of its conservative defenders such as Schmoller, advocated a democratic alternative. Heaving studied under Gierke, he viewed the state as an organic entity, deriving its legal structure from the concretely interacting wills, such as trade unions, of civil society. Legal principles embodied in the constitution were publicly binding, but were created by the interactions of groups of citizens, whose basic rights were defended. In the constitution, not only were individual rights established but also those of community life and economic activity. These views were in accord with Friedrich Naumann, the liberal politician of the empire and author of *Mitteleuropa*, who presided over the constitutional commission on rights. Naumann, whose views on domestic politics reflected those on foreign policy, wanted a socially integrative political system that promoted cross-class unity. He and Preuss followed Max Weber by advocating strong executive leadership, although the resulting presidential powers fell somewhat short of their wishes. The framers' views on the constitution never became definitive, and the viability of the document continued to be vigorously debated. At one pole was the legal positivism of Hans Kelsen, at the other the dictatorial decisionism of Carl Schmitt.

Kelsen, who supported the republic, acknowledged that democracy was inherently pluralistic and that any politics of compromise was incompatible with the notion of absolute truth. The law could be neither deduced from, nor attached to, the wills of individuals or groups. Instead it formed an objective order established by the constitution. The state had no personality and no will as Laband claimed, but was simply part of the legal network. Law itself formed an objective normative order divorced from a metaphysical or subjectivist conceptions. Kelsen's concept of pure law came under attack from a number of fronts, but from none more so than from Carl Schmitt.

Schmitt saw legitimacy not in the system of law but in a unifying will of the political, which he defined as the distinction made between friend and enemy, the self and the threat to the self. This possibility of threat was the basis for political unity, of the people's unified will in the face of the enemy. He argued that the internal divisions in Germany resulted from the supposed neutrality of positivism and liberalism, which had led to a weak bourgeois state. Political parties created a fragmented set of interests that impeded the political will to dominate the enemy. He argued that this unified will could be constituted in "dictatorship," the concentration of power in the face of a threat. In late Weimar, this meant the use of presidential emergency powers at the expense of parliamentary institutions.

During the empire, Georg Jellinek, who was Weber's colleague in Heidelberg, offered a more sociological version of Laband's positivism.

Jellinek believed state must be a *Rechtsstaat*, but added that it is not only a fact of law, but also a social fact. The state as a personality is sociologically established, the product of a complex of "social phenomena." While society can have no unified will, parliamentary institutions in which social forces are represented, can.

With the declaration of the republic, some thinkers viewed the pluralism of civil society as something positive, and joined in a debate over the role of the state and law, their ties to the nation, and the viability of the republic. Gerhard Anschütz, Richard Thoma, and Jellinek's son Walter—all of whom were Alfred Weber's colleagues at Heidelberg—were among those who embraced both—some form of legal positivism and the republic. But I will focus on Hugo Sinzheimer, Hermann Heller, Rudolf Smend, and Gustav Radbruch, each of whom offered a variation of Georg Jellinek's social interpretation of law and the state and tied them to the organic unity of the nation.

Sinzheimer and Heller were Social Democrats who tried to reconcile an organic view of the state with recognition of the importance of civil society. Sinzheimer, who as a member of the constitutional commission was responsible for the inclusion of the right of collective bargaining in that document, wrote that the origins of the state and law were found in the associations of civil society, especially labor unions. If unions and business representatives were allowed to bargain collectively without restraint, they could create the basis for an organic legal administration. This activity was the source of the formation of an organic will and could lead to cross-class consensus. Heller, who studied under Radbruch, also believed that the unity of the state and law emerged from social and economic interactions. The state was an "organic unification of wills" that gave expression to the unity of the history of the nation, its culture and experience. Rather than being a personality above the institutions of society, the state evolved out of them. The state integrated the plurality of wills in civil society into an organically unified will, which became the basis for law. This integration, which was national rather than international, was accomplished through the majority principle and parliamentary representation. Gustav Radbruch, with liberal party affiliations, shared this view of the state. Radbruch arrived in Heidelberg before Alfred Weber, left in the years before the war, and returned in 1926. He attacked the authoritarian state of the empire, arguing that it influenced its subjects through command rather than by educating the citizen. The law of the imperial *Rechtsstaat* lacked any conviction, settling simply for ascetic precision. He opposed legal positivism in favor of a more historical and spontaneous approach to law, whose proclaimed values and duties were embedded in the larger culture. This cultural unity had, as its necessary condition, democracy with the freedom to assemble and form associations. Echoing the ideal of *Bildung*, he wrote that the personalities of citizens were formed in the larger integrating unity, and they expressed subjectively valid legal ideals. Heller, Sinzheimer, and Radbruch assumed that Weimar would remain a stable constitutional system, a belief that was shared by the more conservative Rudolf Smend.

Smend believed that the nation was an organic totality of values that was centered in the religious values of the people. These values could integrate individual citizens into the larger community in a continuous process, in what he termed a "total lived experience." Unlike Sinzheimer and Heller, he did not believe that this integration could take place through parliamentary voting and majority rule but through the plebiscitary authority of the state and its leaders using symbols of the larger set of values. The constitution's role was to give shape to these values and identities, around which citizens organized their lives. The state, thus, was an association of wills into an organic unity of spiritual life. Smend came the closest to Schmitt's emphasis on executive authority, but unlike Schmitt he increased his support for the republic in its later years.

If the last four thinkers had anything in common, it was the conviction that the institutions of the republic were, in some form, compatible with organic cultural values. All of them rejected the contention of the discursive coalition that civil society represented a divisive sphere that had to be presided over by the authoritarian state, embodying the spirit of the nation. This question of exactly how the pluralism of parliamentary democracy could be reconciled with the larger organic culture stood at the heart of Alfred Weber's writings during most of the republic.

Anschütz, Thoma, Radbruch, Sinzheimer, and the Jellineks, all had some connection to Heidelberg. The responses of the Heidelberg faculty were not much different from other German academics regarding the crises of the republic. While the Spirit of Heidelberg continued during the republic, it declined somewhat with the weakening of political stability and material security. Some professors looked back with nostalgia on the prewar spirit and were given to the same cultural pessimism increasingly expressed throughout Germany. Sensing a crisis in their status and ability to speak for a unified public, they lamented the "massification" of cultivation and politics. In addition, they tended to distinguish themselves from those who welcomed modernism.[16] They called for the primacy of spirit in politics and even for a return to notable politics. Like academics throughout Germany, many claimed to be above politics, meaning that they were not active supporters of any political party. However, while elsewhere this stance was accompanied with outright hostility to the republic, in Heidelberg pessimism was not often accompanied by political reaction. While there were certainly reactionaries among the Heidelberg faculty, the university was dominated by professors loyal to the republic or politically neutral.[17]

The ideal of a unified academic cultural perspective informing and inspiring a public sensitized to its leadership by its own academic cultivation appeared largely unattainable without the most dramatic changes, notably in political life. An important question arose: Could a unity be established in conjunction with the new parliamentary state, with its party formations and interest groups, or would it be necessary to revert to something resembling the old monarchical structure? Or put another way: could one democratically extend the ideal of cultivation? And if so, what role were the university and the discipline of sociology to play in this democratization? Many of Alfred Weber's writings at this time were dedicated to bridging this disjunction between "high" and "low" culture. In so doing, they had to come to terms with vitalism in its new Weimar form, especially the work of Oswald Spengler.

Alfred Weber in the republic

With the events of 1917, Weber's immediate concerns switched to the domestic scene. This intensified with the establishment of the republic in 1918. What would the new Germany be like? His primary concern in the prewar period, rethinking Social Policy in light of democratization, continued during the Weimar period, as did his wartime interest in *Mitteleuropa*, although the issue of the relationship between leaders and the democratic masses received the most attention, primarily because of the dissolution of the old discursive coalition. Like his brother, he believed in the institution of the monarchy (if not in its Prussian form), but he came to realize that the empire had so discredited itself that it was beyond preserving. Seeing no use in a ceremonial monarchy such as Britain's, he became a strong supporter of the republic.[18]

He had always had some political ambitions, and the opportunity to realize them came in November 1918, when he was approached by a group of liberal intellectuals to participate in the founding of a new political party, the German Democratic Party (DDP), which claimed to be a bridge between the left wing of bourgeois liberalism and the Social Democrats. Weber accepted the invitation, becoming a leader of the left wing of the party (which included his brother Max) and occupying the party's provisional chair. Although he was effective among the party's founders, none of whom was a professional politician, he proved to be rather incompetent in the larger political process. His chairship lasted less than one month. In December, he accepted rumors that right-wing industrialists Hugo Stinnes and August Thyssen had plotted with the Entente to set up a separatist Rhenish Republic, and he publicly accused them at the DDP's inaugural convention. The rumors proved false, and Weber was forced to resign. After his resignation, the party moved markedly to the right. Later, in December, he ran for the National Assembly, whose primary tasks, to write a constitution for the republic and sign a formal peace treaty with the Entente, did not meet with any success. With his electoral defeat, his formal political activity came largely to a halt, although he remained a party member.[19]

Weber's politics were left-liberal. He believed that the German economy should remain capitalistic but that capitalism had to be reformed along more socialistic lines, allowing for greater economic justice in order to preserve the system. He was not enamored of the SPD, fearing it might engage in radical socialist experiments, but urged the DDP to cooperate with the Socialists in order to act as a moderating force. Stressing the party's mediating potential, he declared that the DDP was a party of the entire people and not just of the bourgeoisie.

> I speak as a so-called intellectual, not a bourgeois. I am a spiritual worker and have nothing to do with the bourgeoisie (Applause). I speak as one of those who had foreseen the decline and collapse of the old.... I want to characterize that which I represent not as bourgeois but as something general, something that personifies the entire people (Lively Applause).[20]

He was also eager to weaken the political position of the former imperial alliance of rye and steel, the noble estate owners and the large industrialists, advocating confiscation of war profits and redistribution of some estate lands among smaller farmers.[21] This position was consistent with his opposition to the old discursive coalition.

During the rest of the republic Weber contributed to attempts to establish a new discursive coalition. He made efforts to continue the

"Spirit of Heidelberg" by fostering the prewar informal institutions. At the beginning of the republic, he helped Max's widow, Marianne, successfully revive the "spiritual sociability" of her husband's circle.[22] This group continued its friendly competition with the George Circle, and Alfred continued to chart a path between them. More important for him were the resumed "Sociological Discussion Evenings," which met every two weeks in a hotel. The lively debates that took place there attracted a number of rebellious young intellectuals. In addition to these informal gatherings, Weber continued to participate in existing formal institutions such as the German Sociological Association. Nonetheless, such efforts did not keep the Spirit of Heidelberg from declining during these years.[23]

In pursuit of his organic cultural, domestic, political, and foreign policy agendas, Weber participated in the founding of three new institutions: the Institute for Social and State Sciences, the Weimar Circle, and the European Cultural League.

In 1923, with the retirement of his colleague Eberhard Gothein, Weber took over leadership of the National Economic Institute at the University of Heidelberg. The next year, he changed the name to the Institute for Social and State Sciences (InSoSta) in order to indicate the importance of sociology for the state sciences. That same year he insured that a new, privately funded chair of foreign policy in the Institute be occupied by his student Edgar Salin, a specialist in Finance and a member of the George Circle. The InSoSta was one of a number of similar institutes founded during the republic, including the Academy for Labor in Frankfurt, the College for Politics in Berlin, and the Institute for Social Research in Frankfurt. All of these, to varying degrees, shared three objectives: to educate students in accordance with the new reform pedagogy that had developed out of the youth movement, to promote a democratic orientation, and to overcome the isolation that characterized German science in the wake of the war. The InSoSta was less of a research institute than the others, at least at its beginning.[24]

The Institute, dubbed by one scholar as the "Institute of Outsiders," included, among its members, those who were underrepresented in German universities. Over half of the lecturers were of Jewish background, and over half belonged to democratic parties like the DDP and the SPD. A large number of the students came out of the youth movement.[25] Among those who participated in the Institute either as professors or students were Edgar Jaffé, Hans Gerth, Emil Lederer, Karl Mannheim, and Norbert Elias.[26] The latter three would develop their own versions of cultural sociology. In keeping with the Spirit of Heidelberg, Weber tolerated the entire spectrum of political positions. Jakob Marschak was a Menshevik; Arthur Salz was involved with the Munich Soviet Republic; Arnold Bergstraesser developed ties to the reactionary general and chancellor Kurt von Schleicher; Ernst Wilhelm Eschmann became an admirer of Mussolini's fascism and, with another InSoSta student Giselher Wirsing, assumed editorial leadership of the right-wing journal, *Die Tat*. Weber maintained his ties to all of them despite his personal dislike for many of their political positions.[27]

To maintain this atmosphere of tolerant intellectual exchange, Weber insisted that the Institute remain dedicated to the cultivation of the German elites and resist purely instrumental instruction.[28] This corresponded to his emphasis on the need for social scientists to provide cultural orientation for those who would become the leaders of Germany. He left the teaching of national economy to others as he concentrated on cultural sociology. The work he considered his most important, *Cultural History as Cultural Sociology*, originated in lectures he delivered at the Institute after 1926.[29]

In the mid-1920s, Weber became a leader of the Weimar Circle, a group of academics who actively supported the republic. While the correlation between this group and those who challenged the discursive coalition in the empire was not exact, it was roughly approximate. Organized by the historian Friedrich Meinecke, most of its members were historians, political scientists, and national economists (sociology as a university discipline had a minimal presence at this time) who had opposed imperial annexation during the war. Of the legal theorists discussed above, Anschütz, Sinzheimer, Thoma, Radbruch, and Walter Jellinek attended at least one of the Circle's three conferences.[30]

The Weimar Circle represented a spectrum of positions ranging from rational republicanism to democratic idealism, Weber being at the latter pole. Most members, like Weber, were associated with the DDP. Because professors who supported the republic tended to be more passive than those who opposed it, the Weimar Circle sought not to politicize academics, but just the opposite, to discourage academics associated with the right-liberal German People's Party (DVP) from political activity in opposition to the republic, and encourage the same loyalty to the state these academics had practiced during the empire.[31] In short, they hoped to create a new discursive coalition.

The Weimar Circle recognized that democracy was an inevitable part of modernization. Most viewed democracy not as a sociophilosophical goal but as the activation of those strata that had been inactive before the industrial revolution. Most separated the concept of democracy from that of parliamentarization. They increasingly believed that the latter should be limited through a stronger executive in order to protect the former. The relationship of the masses to the elites, a widespread concern in the republic, was central to their concerns, as was Weber's theory of "leader democracy."[32]

In 1925, Weber was invited to join the European Cultural League, which had been founded three years previously in Vienna. In 1926, a German section of the Cultural League was founded in Heidelberg, and Weber was elected its vice president, a position he held until 1933. A number of InSoSta members participated in the League, and one of Weber's students, Max Clauss, became the editor of the League's journal, the *Europäische Revue*. The purpose of the League was to bring European intellectuals together and to promote a "common European consciousness" in the journal, a goal very much in accord with Weber's own.[33]

Although the League offered Weber the opportunity to publicize his political views about Europe—he gave presentations at the meetings and published articles in the journal—he did not set its agenda. And like the InSoSta, the League contained a wide variety of political positions. But most of the German members, including Weber, shared two basic principles: they believed in the ideal of a spiritual aristocracy, and they wanted Germany to have its rightful place among the European powers, that is, they wanted a revision of the Versailles Treaty. Nevertheless, the group tended to be more conservative than Weber, combining an antidemocratic elitism with a desire for French-German cooperation. Clauss himself had ties to the conservative *Tat* Circle and, in opposition to his mentor, quickly came to support the Nazis. His belief that Germany's task was to become the center of gravity of the continent was a more expansive version of Weber's idea of *Mitteleuropa*.[34]

These three organizations—the InSoSta, the Weimar Circle, and the European Cultural League—embody Weber's ideals during the republic. He wanted the InSoSta to cultivate the next generation of German leaders in a way that connected them to the organic values of the nation. This meant making cultural sociology a central part of the curriculum to balance the specialized training that students would receive. The Weimar Circle represented his wish that the spiritual aristocracy in the university not be divorced from the democratic political process of the republic. And the European Cultural League promoted the same kind of unity between the elites and the larger masses throughout Europe that the Weimar Circle hoped to achieve in Germany. Its German members sought to establish a European community without sacrificing German national identity.

Weber's participation in these institutions could be conducted with some degree of optimism. The republic had weathered its turbulent early years and had settled into five years of relative calm. The "Golden Twenties" witnessed a burgeoning of cultural activity of various sorts, and the political middle held. That changed with the onset of the world depression in late 1929. The break-up of the Weimar Coalition in March 1930 over the financing of unemployment benefits resulted in parliamentary gridlock for the remainder of the republic's short life span. If Weber's formal political activity was halted in the first years of the republic, his cultural politics suffered the same fate with the demise of the republic in 1933. The problems began in 1930, when Karl Mannheim left to accept a chair in Frankfurt, taking his students, including Norbert Elias and Hans Gerth, with him. The next year, Emil Lederer left for a chair in Berlin. They represented two of the most important members of the left wing of the institute. Their replacements came from the right wing, which now dominated the institute. Some of these members had clear fascist leanings and most cooperated with the Nazis after they came to power.[35]

With the victory of the Nazis and their Nationalist allies in the March 1933 election, the two parties began to replace the flag of the republic on public buildings with their own party flags. When the Nazi flag went up on Weber's InSoSta, he ordered the director of the Heidelberg police to take it down, a command that was obeyed. But the director was soon replaced, and the flag went up again, this time protected by an armed guard. Weber briefly closed his institute in protest. When one of his colleagues, Hans von Eckardt, was dismissed from the InSoSta despite Weber's vehement protest, Weber understood that the university was in the process of being nazified. In April, he requested that he be retired to emeritus status, a request that was granted. Most of the remaining members of the "Institute of Outsiders" were dismissed, beginning with those who were Jewish. The Cultural Sociological Seminar and the Sociological Discussion Evenings came to an end. Weber's entire cultural sociological establishment had been dismantled, and he entered into the "internal emigration" of many German intellectuals.[36] The Spirit of Heidelberg had clearly expired.

7
The Sociology of Culture in Weimar

Alfred Weber did not return to the formulation of the sociology of culture until after the end of the war. In so doing, he had to contend with a set of new intellectual forces, beginning with the views of his brother Max. At the beginning of the republic, two lectures that Max had given to students in Munich on politics and science as vocations, or callings, appeared in print. Both lectures challenged not only important premises of the discursive coalition but also some of Alfred's own assumptions. The expansion of Max's *Protestant Ethic* into three volumes on world religions and the posthumous appearance of his monumental *Economy and Society*, together with his earlier works, provided a modernist interpretation of culture, a methodology, and political philosophy that emphasized the individual, and a metahistory that focused on the process of rationalization within which the individual had to navigate.

A second challenge came from Marxism, which assumed a more prominent place in the republic. Georg Lukács's *History and Class Consciousness*, which appeared early in 1923, was especially important. Lukács had been a participant in Max Weber's circle before the war, and during the republic, his erstwhile protégé Karl Mannheim took up residence in Heidelberg, under Alfred Weber's sponsorship, using the cultural capital of his earlier association. Weber never addressed Lukács's book directly, although he did hold a joint seminar on it with Mannheim. In that seminar, the focus was not on Lukács but on Mannheim, whom Weber accused of practicing a sophisticated version of Marxism. This confrontation will be examined in the next chapter.

Ferdinand Tönnies's *Gemeinschaft und Gesellschaft*, which had been so constitutive for Weber's generation, became a favorite of many who looked nostalgically back at the empire. These people believed that the republic, which Tönnies supported, represented the decline of the national community. The book realized its height of popularity during the first half of the republic.[1] Even more popular was another pessimistically dualistic work, Oswald Spengler's *The Decline of the West*, which offered a gloomy antirepublican interpretation of the fortunes of the modern Western world. Weber implicitly changed the successional model of both of these books, and he explicitly questioned Spengler's.

Oswald Spengler and cultural pessimism

The downfall of the imperial state and the turmoil of the early years of the republic resulted in considerable disorientation among many steeped in the habitus of the imperial discursive coalition. Not only academics, but also many within the larger cultivated public viewed the democratization of culture with trepidation. Faced with a crisis of classical modernity (even though they might not have expressed it as such), disenchanted with mainstream science, they sought an antimodern solution, something more intuitive and even mystical. The rigidity of academic specialization provoked unrest in German youth and the desire for some kind of spiritual revolution, a new synthesis, and a new ideal of cultivation.[2]

Vitalism, which during the empire had significant support among countercultural elements and even resonated with some (such as Alfred Weber) within the academic community, enjoyed increased support within the cultivated public after the war.[3] Ernst Troeltsch, whose organic historicism was akin to Weber's, saw vitalism contributing the new methodology for a "revolution in science." He listed, as the leaders of this postwar movement, members of the George Circle, especially Alfred Weber's associate Friedrich Gundolf,[4] and Oswald Spengler, whose book Troeltsch saw as the first great breakthrough for the new science in public opinion. He believed that cultural movement could only be understood through the methodology of Gundolf and Spengler.[5]

The extent of the cultivated public's willingness to accept a synthetic vitalistic outlook can be seen in the enormous popularity of Spengler's *The Decline of the West*. The first volume of the book appeared in the summer of 1918, shortly before the end of the war and the creation of the republic, and sold out in six months. Within eight years it had gone through fifty-five printings and sold a hundred thousand copies.[6] Ernst Cassirer described its impact:

Perhaps never before had a philosophical book such a sensational success. It was translated into almost every language and read by all sorts of readers—philosophers and scientists, historians and politicians, students and scholars, tradesmen and the man in the street.[7]

There were a number of reasons for the book's success. Cassirer pointed to the book's pessimistic title, which seemed appropriate in the wake of the war, and to its almost mythic quality, analogous to astrological treatises of the time.[8] Widdig notes that the book took its structure from popular dichotomies of the time that "resonated strongly within the ranks of an insecure and beaten German middle class." The boldness and authoritative tone of the book's uncritical generalizations were supported by detailed empirical material.[9] The very bulk of the book, rather than discouraging readers, offered an aura of certainty. The general public could assume that the large amount of detail was "proof" of the repetitious message.

Specialists attacked the book after its appearance, but had little impact on its popularity. And, within a few years, some of the criticism had yielded to grudging admiration.[10] Here was an outsider who accomplished what had become increasingly difficult for the academic defenders of the discursive coalition during the later years of the empire. He offered a diagnosis of the time that appealed to the cultivated public and *appeared* to retain the scholarly rigor of the specialist. When Alfred Weber offered his own diagnosis based on cultural sociology, he would have to come to terms with this cultural metahistory, which used many of the same binary constructions he used. Both men offered a vitalistic view of the cultural half of their central culture-civilization dualism. Both emphasized the importance of leadership, even if the nature of that leadership differed markedly. Both sought an audience among the larger cultivated public. And both would make a synthetic metahistory a central part of that appeal.

As the foremost example of Weimar vitalism,[11] Spengler's world history illustrates a set of universal principles via the histories of the "higher" cultures, the classical, the Western, the Arabian, the Egyptian, the Indian, the Chinese, and the Mexican. Although all of these are discussed to some degree, the first three, and especially the first two, draw the most attention. To a large extent, the book focuses on the contrast between the guiding principles of the classical and Western cultures, but that is of less importance to us (and probably to Spengler's Weimar readers) than his use of this contrast to prophesize the decline of the West. The book is as culturally relativistic as mainstream German historicism insofar as it portrays each of these cultures as a unique entity that cannot be judged by the standards of another one.[12] However, all of the cultures developed according to the same analogous pattern—the cycle of the four seasons. Given this pattern, there was no way that the current reigning culture, the West, could escape its destiny as it descended into its "winter."

The idea of destiny not only characterized one side of Spengler's initial dichotomy, history and nature,[13] but also informed all of his other binary constructs, which included: organic-mechanistic, aristocratic-democratic, estates-parties, country-city, people-masses, community-society, state-economy, soul-intellect and culture-civilization. The first term of each pair was the one privileged by Spengler, just as it had been by the articulators of the discursive coalition. In fact, there was nothing about Spengler's definitions of the terms that would have challenged the coalition's conservatism—with the important exception of the last pair, culture-civilization.

That dichotomy, so important for Alfred Weber, organized the others in a way more akin to the challenge of Weber's generation of national economists to the imperial discursive coalition in that it added a temporal element just as Tönnies's *Gemeinschaft-Gesellschaft* dualism did. Rather than view civilization as simply the opposite of culture, Spengler saw it as the last stage of culture. When a higher culture entered the winter of its development, it lost its vitality and became a sclerotic civilization. In civilizations, the second term of the above-listed binary constructions dominated—for example, democracy replaced aristocracy.

Spengler's treatment of the relationship of the state to the economy within this temporal binary construction is instructive. His discussion was grounded in the vitalist concept of "race," which, he argued, is characterized by "blood." Through the vehicle of language, a race achieves waking consciousness in the form of the soul. A "people" (*Volk*) is a unit of the soul of the culture, and when it achieves a certain historical presence, it becomes a nation, the largest unit in the "stream of history." Only higher cultures achieve nationhood. The state is the highest law-making subject of the nation, and, at this stage, is a cognate of the nobility.[14] In these definitions, Spengler adopted privileged terms of the imperial discursive coalition—people, culture, estate (*Stand*), nobility and state—and set them against corresponding negative terms—masses, civilization, party, democracy, and economy—as the discursive coalition did. However, by linking these dichotomies to the cyclically determined idea of race, he necessarily arrived at a pessimistic conclusion. While the state may represent the historical zenith of a culture/people, it must give way to the civilizational stage dominated by the money economy. Blood is overcome by another circulatory substance, money.[15]

Spengler chose a strategy similar to that of Alfred Weber's generation of the Historical School, who challenged their predecessors by turning the hierarchical dualism of state and civil society into a successional one. However, he differed from them in several ways. Unlike Tönnies, who offered a set of ideal types for the analysis of specific institutions, Spengler applied his chronological dichotomy to entire cultures and offered no hope for the revival of Western culture. Most importantly, he did not assign the state completely to the civilizational side of the dualism. In essence, he depicted two states, a cultural institution (embodying aristocracy, blood, and soul) and a civilizational institution (governed by democracy, money, and pure intellect). His work represented a nostalgic reaffirmation of the most reactionary formulation of the old discursive coalition, while admitting that it could never return, and a condemnation of both the parliamentary democratic forces that Weber's generation supported and their hope to establish a new democratic discursive coalition.

Spengler did offer some hope—although not for Western culture—in the form of Caesarism. He wrote that when a culture reaches the final chaos of its civilizational-democratic phase, it can be overthrown by Caesaristic might. This triumph will mark the transition to a new culture governed by a new aristocracy. Once again, blood will have triumphed over money.[16] This political message at the end of his book became garbled in a series of short political statements in the mid-twenties. However, what is clear is that he attacked the importance of "cultivation and the cosmopolitan bourgeoisie and the spiritual mission of the Germans."[17] And he reiterated his intense dislike for parliamentary democracy.

There were elements of Spengler's work that Alfred Weber admired—the vitalism, the emphasis on culture, and his ability to address a larger audience. But he abhorred Spengler's political position, which could not be divorced from his cultural views.[18] This discrepancy meant that Weber had to be very cognizant of Spengler as he developed his own cultural and political ideas.

Weber's sociology of culture in Weimar

Weber's cultural writings during the republic were both an extension of his prewar work and an attempt to modify them into a more precise form, while attempting to steer a course between the fragmentation he perceived in the republic and Spengler's pessimistic alternative. These writings on the sociology of culture are of two types. The first type[19] discusses the nature and place of the discipline in a manner akin to his 1912 presentation to the German Sociological Association. The second type[20] is dedicated to "cultural history as cultural sociology," that is, it offers a metahistory to compete with already-existing ones, especially Spengler's.[21] In the first type of writings, Weber repeated the binary structure of the prewar talk, but with some significant modifications. These were designed to address the critiques of his earlier statement, to distinguish his sociology of culture from its competitors, and to address the changed circumstances in Germany. He believed that his approach best offered a sociological diagnosis of "our place (*Ort*) within the general movement of history."[22]

"Our place" became a more defined entity in the Weimar writings, a spatial and temporal whole that was alternately designated as a "historical circle" (*Geschichtskreis*) or a "historical body" (*Geschichtskörper*). Although Weber's use of the terminology was inconsistent, the former generally refers to larger units, such as the ancient Near East, and the latter to subunits, such as ancient Egypt and Babylonia.[23] Historical bodies drew most of his attention. The concept allowed him to avoid simply having two general types of civilization and culture within the historical stream. Now he posited a more delineated entity in which those types and a third type restored from the 1909–1910 lectures, social process, came together in unique configurations (*Gestaltungen*). Cultural sociology was charged with the analysis of these configurations. Weber, like Spengler, used vitalistic biological analogies, describing his approach as "morphological," but despite indications of the general similarity, he was quick to disassociate his use of the term from Spengler's.[24]

In depicting the historical body, Weber turned to Driesch's concept of entelechy over Bergson's more amorphous life-stream. Driesch had concerned himself with specific organisms and not with the unity of all life (although he paid lip-service to it).[25] For Driesch, individual parts of an organism develop in accordance with the entelechy of the entire organism. These parts are different from one another, but all embody the larger pattern that gives the organism its identity. Weber took the same approach. The parts of a historical body, that is, individuals, might be themselves unique, but they are part of a whole, with its own distinct pattern. In 1931, he actually used the term "entelechy" and drew the analogy to biology,[26] but in the 1920s he used the term "community of destiny" (*Schicksalgemeinschaft*).

Weber believed that different spheres of human activity interact in historical bodies and that it was the task of the sociologist to typologize these spheres. In keeping with his organic approach, he warned that the division of these spheres into types was simply for analytical purposes, for in actuality they do not exist independently of one another. In Weimar, a third type, the "social process," was added to the two prewar types of civilizational process and cultural movement. Volker Kruse writes correctly that the new structure of types is based on the traditional triad of body, spirit (*Geist*) and soul (*Seele*). Nevertheless, I believe it is more rewarding to view the triad as mediating two binary constructions, larger natural-human (body-spirit / soul) dualism and, within the latter, that of spirit-soul.[27] Weber's lack of sustained attention to the social process supports this view.[28] He never really broke free of his earlier binary orientation.

His description of the civilizational process as a "process of rationalization" remained largely unchanged from his prewar formulation. Aimed at the observation and manipulation of both internal and external existence, the mastery of nature, this process was largely utilitarian. It incorporated all abstract theories of nature, from Euclidean geometry to the Copernican theory, as well as technological advances such as the steam engine. Because it proceeded according to the logic of causality, both its generalities and their parts were "discovered" rather than "created." It followed its own pattern of development, one of universal linear progress,[29] which was described optimistically by the thinkers of the Enlightenment as the idea of progress and, more pessimistically by thinkers like his brother Max, as the rationalization process. In the republic, Weber modified his use of the term "spirit" (*Geist*). Earlier it had been used to refer to noncorporeal human attributes in general; now it applied only to those attributes that are rational, or "intellectual." It referred to that aspect of the mind that can be conceptualized, externalized, and objectified. This potential for objectification allows it to be a communicative medium with other individuals and, more importantly, other historical bodies. In short, Weber used "spirit" primarily to describe the civilizational process.[30]

As a complement to the civilizational process, Weber restored the type from the 1909–1910 lectures that had been dropped in 1912, the social process. While this new type certainly makes his account appear more sociological, he in fact did very little with it. He defined the social process as the social objectification of unchanging natural forces of drives and will, which, in turn, are given pattern by certain natural conditions such as geography and climate.[31] Historians might ignore these patterns and present the facts organized by them as the events of politics, the economy, and society, but sociologists focused on the patterns as typical forms and stages of development. These stages, which Weber described as "inherent," "predetermined," and "necessary," are essentially those of the body, moving from youth to old age, the final stage being "senile torpor." Just as individual biological bodies, while sharing a basic life cycle, vary in their physiognomies and their ability to remain vital, so do historical bodies. Some descend into lasting rigidity more quickly than others. Weber also agreed that great individuals influence the life cycle, as do a body's specific geology and climate, but there was no specific account of how this occurs. Even though he referred to this type as a "process," its pattern is different from the linear one of the civilizational process. It is cyclical and, thus, not at all historical.[32]

Although Weber used the same pattern present in Spengler's account of cultures, he did not use the latter's seasonal analogy. More importantly, he indirectly acknowledged Spengler's cyclical model while shunting it out of the spiritual-cultural sphere and into that of society, where it was joined with materialism. Whether it was his intent or not, Weber's restoration of the social process allowed him to acknowledge Spengler and incorporate his cyclical approach without allowing it to play a primary role in his cultural sociology. That role was reserved for the other two types of civilization and culture.

Although they have different patterns of movement, cyclical and linear, the social and civilizational processes are closely related.[33] Yet Weber clearly privileges the latter. A historical body's place within this process is much more important for its unique identity than is the stage of its social cycle. And the civilizational process provides historical bodies with the technical means for their forms and social structures. Despite differences in their social processes, these bodies build on one another within the unilinear civilizational process. This process might become stalled in a certain historical body due to its social structure, but the most important civilizational elements will be taken up by new historical bodies. When the interaction of the two processes within a historical body results in significant change, a new "life-aggregation" appears. It then becomes necessary to make sense of that aggregation, and that task falls to culture.

The civilizational and social processes help shape the historical body, but they do not give it its entelechial essence. That can be found only in the sphere of culture. This sphere remains the most important for Weber and, despite his increased explication, the most evasive. In 1912, Weber labeled culture a "process," just as he had labeled civilization. Now he reserved the term "movement" to distinguish culture from both the civilizational and social processes. He also employed a relatively new term for him, the "soul" (*Seele*), to describe culture.[34] The destiny, that is, entelechy, of a historical body is expressed through its soul. With this term, Weber attempted to avoid the amorphousness of the earlier "life feeling." "Soul" refers to the inner unity of a person, an essence that precedes acts of thinking, feeling, and willing, and is perceived as a gestalt, an organic whole that cannot be analyzed through its parts. This unified entity is monadically at one with the larger soul of the historical body, the totality of meaning and values. Its resistance to rational analysis made "soul" a favored term with vitalists and others who, in the words of Peter Gay, had "a hunger for wholeness."[35] It was certainly prominent in Spengler's work.

Weber wrote that the soul aspires to achieve "redemption" (*Erlösung*), by which he meant a unity of meaning in the face of the realignments brought about by the civilizational and social processes. This aspiration (*Wollen*) to form the stuff of existence in accord with its own essence is realized through the creative cultural act. He defines culture as the "attempt of the soul to achieve the expression, gestalt, image, and form of its own essence."[36] If the soul of the historical body seeks unity, its gestalt, then the creative act of culture is a "configuration" (*Gestaltung*) of the symbols through which the soul expresses itself. Thus, the redemption of the soul is the achievement of oneness with the (undefined) "absolute" through a cultural configuration. Two questions remain: What is the extent of this unity and what role do individuals play in its creation?

Weber's answer to the first of these questions is very much akin to that of Driesch. The entelechy represented by the soul is confined to the historical body. Whether the "absolute" is a larger unity beyond historical bodies, as was Bergson's life stream, is not important to him. Like Spengler's, Weber's discussion was relativistic—the unity that the culture of a historical body achieves is not transferable beyond that body. This differentiates cultural movement from the civilizational process, which does extend beyond individual historical bodies. Cultural movement does not develop linearly as does the civilizational process; rather it is characterized by disconnected periods of productivity, "new psychic (*seelisch*) situations, a rough sea, sometimes cresting sometimes calm, moved by this or that 'psychic' wind, but never 'flowing' to somewhere, never striving toward a 'goal.'"[37] These changing situations are largely the result of the social and civilizational processes. In some cases they bypass individual historical bodies, leaving the soul and its cultural expression unchanged. In other words, they significantly impact those bodies, necessitating that the resulting new life-aggregations be brought into accord with the soul of the historical body. Such situations demand essentially a new cultural configuration.

Where does such a configuration come from? Weber answered that it occurs through the creative acts of individuals. Great men, those creative individuals who respond to the disruption of the life-aggregation by bringing the stuff of existence into line with the soul, their own and that of the larger culture, are the standard bearers of any culture. Weber's monadic formulation here echoes his earlier comments, in response to Hans Staudinger's book, that the oppositional dualism of individual-community is a false one. Communities are not the result of the repression of personality but rather their meaning and values are expressed through personalities.

Weber believed that when this creative endeavor is stymied, a crisis of culture results. Like most who perceived this cultural crisis in the republic, he saw it taking the form of relativism, which he defined as the inability of the soul to configure itself. The current crisis was due to the overvaluation of spirit in the nineteenth century, in part due to the significant technical mastery of nature and the spread of Enlightenment rationalism. Knowledge was sought simply in the sphere of spirit, and history became a teleological process. Thinkers such as Hegel, Marx, and Lukács viewed history as a rational process and largely conflated the spirit and the soul to the detriment of the latter. Whether they placed emphasis on the spirit of the civilizational process or on the social process, cultural movement was largely ignored. As a result of this "intellectualism," people perceive an apparent dissolution of the "contemporary psychic (*seelisch*) configuration." With the old forms of psychic expression appearing empty and meaningless, the absolute appears relativized. Unable to understand the meaning of the world in its totality, people seek any path to redemption.[38]

Weber feared a drift toward extremes such as "intellectualistic" positivism and orthodox Marxism on the one hand and irrational ones such as the mysticism of a Stefan George on the other, the latter being an extreme reaction to the former. The cultural pessimism that reads Tönnies's and Spengler's books simply as narratives of decline is a product of the rationalization process. Like his brother Max, Weber sought a more constructive alternative, which, he believed, would be informed by cultural sociology. Understanding the relationship of the cultural soul to the other human activities, the discipline brings a realization that every new "life-aggregation" presents the possibility of a new psychic configuration. The viability of such a discipline is tied to its methodology.

The methodology of cultural sociology

Although Weber's 1920 formulation of cultural sociology remains the most extensive discussion of its three types, it did little to address the question of methodology. He sought to rectify this omission in the introduction, titled "Tasks and Methods," to his 1927 collection of earlier essays in cultural sociology, which meant, above all, distinguishing his approach from that of his brother Max. He specifically outlined two basic methodological differences between the two.[39] First, Max, in order to achieve a conceptual exactness, sacrificed any attempt to grasp phenomena as belonging to a totality. He acknowledged the great value of Max's ideal-typological "apparatus" in presenting valuable conceptual schemata retaining an "abundance of conceivable or real causal connections between the parts of the totality of life." However, this emphasis on pure conceptual formation through exaggeration and isolation results in "a net meshed somewhat too widely for the accommodation of the material if one introduces the concepts for the characterization of something that is entirely concrete." Second, he wrote, Max's approach considered social structure and its movement as well as all transpersonal forces strictly through "the social intentions, attitudes and reactions of individuals." Such an approach divorces the parts (individuals) from the larger organic totality that culture represents. Alfred proposed an alternative:

> We want to illuminate complex totalities in their complexity while consciously preserving them as unities, since our whole intent is directed to the passable understanding of insoluble historical collectivities that at their core are quite irrational in their *unity*. It is directed to large, assembled total phenomena, which are traversed by infinitely many individual causal series of the most *heterogeneous* kind that are of no concern to us.[40]

These two methodologies take different approaches to empirical studies. Three aspects of Max's methodology noteworthy here are: the emphasis on an individualizing heterogeneous causality; the concept of understanding (*Verstehen*); and the main form of conceptual formation, the ideal type.[41] In 1904, Max wrote: "'Culture' is a finite segment of the meaningless infinity of the world process (*Geschehen*), a segment on which human beings confer meaning and significance."[42] When human beings act, they do so in a world that makes sense to them according to the meaning that they have given it. Because action is ultimately individual, so is the meaning. However, the action takes place in a social context. If individuals decide on a certain course, they have to anticipate how others will respond to that course. They try to make sense of others' actions, just as cultural scientists might try to make sense of theirs. This position has an important corollary: it assumes that one can make sense of all *social* action. All understanding is, to some degree, rational.

To generalize about this individual social action, Max devised the "ideal type," which he described as a one-sided *accentuation* of "one or more points of view," a unified *analytical* construct," that "in its conceptual purity" cannot be found empirically anywhere in reality. It is a *utopia*."[43] As such, it is not a hypothesis but a heuristic instrument to aid in the construction of specific hypotheses. In *Economy and Society*, he listed four ideal types orientation for social action: instrumentally rational (*zweckrational*), value-rational (*wertrational*), affectual, and traditional.[44] Because these types seldom, if ever, occur in their pure forms, most social action results from some mixture of them. And it may be simply be that what the individual perceives as orienting others is actually his or her primary motivating factor.[45] A social relationship then involves a "plurality of actors" who take account of the orientations of one another. This relationship can be one of cooperation or conflict.[46]

Max constructed two ideal types of social relationships. When the relationship is based on affectual or traditional orientation, he termed it "communal" (*Vergemeinschaftung*). When it is based on value-rational or instrumentally rational orientation, it is "associative" (*Vergesellschaftung*).[47] Like the two basic social relationships, the myriad of types that Max introduced in *Economy and Society* derive from his four ideal types of action. All are, of course, themselves ideal types. For example, he characterized organizations by their types of "domination" (*Herrschaft*), or authority, which, in turn, depend on the legitimacy in the eyes of both—those who issue orders and those who follow them. The three "pure" types of domination/authority he identified are legal (based on value-rational and/or instrumentally rational grounds), traditional (based on traditional grounds) and charismatic (based on charismatic grounds, themselves a form of affectual orientation).[48] Thus, all larger structural forms are built out of components consisting, basically, of individual action.

It was to this composition that Alfred objected. He described the stages of his methodology:

> Total intuition of the stuff of historical existence of a time, dissection and classification of the same according to the large indicated categories or at least with reference to these, illumination of the same apart from the thusly analyzed total constellation and then the reinsertion in that improved total view corrected by the specialized analysis.[49]

If Max started with the individual and worked out, Alfred started with the general and worked in. The cultural sociologist interprets the interaction of the civilizational and social processes and cultural movement in "the completely concrete, unique constellation of a historical moment."[50] He described the method of the cultural sociologist as akin to phenomenological "intuition" (*Schau*), but differing in that it was not only intuitionistic and synthetic, but also consciously analytic. Analysis of the cultural sphere consists of clarifying how "the always creative, psychic human power" is located in the "life substance" according to its social, civilizational, and cultural attributes. It investigates the "sociological constellations and their transformations in the historical process" to clarify their pattern and conditions. It does not attempt to arrive at "an exhaustive causal explanation."[51] With the exception of the three forms of historical movement, Alfred never offered the detailed taxonomy of types that Max did in *Economy and Society*. His method can only be made more specific by examining its application.

The best example of Weber's historical sociology of culture, "Cultural-sociological Attempts: The Ancient Egyptians and Babylonians," appeared the year before his methodological explication.[52] I will discuss only his analysis of Egypt, since Babylon was treated as simply just an extension of it. He sought to demonstrate that the Egyptian world, rather than being narrowly rigid and unchanging, displayed a whole range of new elements similar to that of modern psychic life. However, the many innovations were simply new expressions of the same unchanging structure, which he referred to as "productive repetition."[53] The two questions he seeks to answer are: How did such the Egyptian structure arise and why did it not change until the time of Akhenaten? To answer these questions, he wrote, one had to examine the interaction of the civilizational and social processes with the cultural aspiration.

Weber argued that when the Old Kingdom was established over the earlier tribal societies, a unique sociological constellation was established, which would last a millennium. He pointed to the construction of a canal system by the early pharaohs that exploited the regularities of nature such as the weather and the Nile River flooding to create a stable agricultural economy and a nonnomadic farming people. The technological achievement of the canals is assigned by Weber to the civilizational process, as are other innovations that were grouped around the canals (such as the calendar and writing). These developments allowed for a centralization of power and the growth of a rational bureaucracy. The bureaucratic apparatus with its "hierarchy of scribes" became the center of the new system. When new social strata arose, such as officer corps and priests, they were integrated into the bureaucratic system.[54]

To this point the connection of the civilizational and social processes is not that different from a Marxist depiction, but Weber took pains to separate himself from historical materialism by adding the cultural dimension as something more than mere superstructure. He argued that the religion of the earlier tribal society had been a totemism of nature gods. As centralization took place, the old habitus of local gods was also centralized with the pharaoh as the central god. He was identified with the canal system as the other gods were identified with aspects of nature. The result was a "marriage" of high civilization (and bureaucracy) and old primitive religiosity, resulting in a new form of cultural expression, the visible form of which was constancy. The progress usually associated with the civilizational process was arrested and innovations became simply more refined ways of repeating the old form.[55]

When systemic change did come, Weber argued, it did not result from an "enlightenment" or a weakening of ties to the primitive religion. Rather, it was the innovation of the Pharaoh Akhenaten, for only the pharaoh-god who presided over the religion and the bureaucracy could bring about such change. He established a monotheistic cult, elevating a heretofore minor god to the supreme position and associating himself with the new deity. Weber believed this religious uniformity was designed to establish Akhenaten's domination beyond Egypt proper to the rest of the empire, a task for which the old religious forms were inadequate. In doing so, he shattered the traditional "life-aggregation" that had survived for so long. "It did not die from the weakness of old age, but was murdered by history."[56]

This empirical study shows the similarities and differences between the methodologies of the Weber brothers. Both approaches reflect the training they had in the Historical School of National Economy. Both denied that cultural sociology could be nomothetic, and both understood that the purpose of generalizations was the illumination of specific historical situations. Alfred never did consider himself to be constructing ideal types, but he did understand that the interaction of his three types of movement was unique in every situation. Where the brothers differed was over the matter of the relationship of the individual to the larger society. Max's approach was based on the individual. Larger units were simply assemblages of their individual parts. Accordingly, his concepts of understanding (*Verstehen*) sought to make rational sense of individual motivations and action. This belief determined the types he constructed. Alfred, on the other hand, believed that the unified vitalistic force of "life" flows through all human existence. Larger units and the individuals who compose them exist in a monadic organic relationship. This assumption he shared with the older concept of cultivation (*Bildung*). Accordingly his types were those of more general historical movement that characterize the three spheres of society, spirit, and soul, all of which are both individual and collective.

Cultural history as cultural sociology

In 1931 Weber wrote an article on cultural sociology for a compendium on German sociology featuring most of the important names in the discipline. This piece would sum up his previous work and set the tone for his succeeding writings. In addition to reiterating his model of the three spheres of culture, civilization, and society, he added some new terminology (without really changing the message), and he pointed to a more historical approach along the lines of his study of the ancient Egyptians and Babylonians.

This essay established the basic terminology for the works that followed, as new terms were integrated with previous ones. Making its debut was "immanent transcendence,"[57] which appears to be roughly identical to his earlier Bergsonian "life forces": forces that are beyond our full powers of perception, yet impact our lives and thus are personally experienced and intuited.[58] Weber was concerned with the specific configurations of these forces in history, and he now adopted Driesch's term "entelechy" to discuss such a "cultural physiognomy" of a "historical body." Rather than being biological like Driesch's, Weber's entelechy was "psychic" (*seelisch*), indicating that the "soul" provides an organic unity for a historical body that is greater than its parts.[59] Even if the parts fragment, the unity of the whole remains. To discuss the interaction of this soul-centered culture with the "life-aggregation" of the social and civilizational spheres within a historical body, Weber had introduced the concept of "constellation" sporadically before the war, gave it a central place in 1927, and he continued it here.[60] Despite the modification of his terminology, the goal of Weber's cultural sociology remained the depiction of the three spheres within specific historical settings.

The most important of these settings for Weber was his current one, and thus the main task of cultural sociology was to answer the question: "where do we find ourselves in the stream of history?"[61] In answering this question, he found it necessary to distinguish his approach to history from others that privileged the civilizational sphere over the cultural one. Whether it was the Hegelian philosophy of history, an evolutionary positivistic approach, or Marxian historical materialism, a priori categories were imposed on the totality of history in a teleological manner, all of which privileged the civilizational sphere in one way or another. Spengler's morphological approach, to the contrary, focused on the soul of culture, but did so by viewing culture simply as the development of the soul according to the prescribed cyclical pattern of the seasons, ignoring its interaction with the other two spheres.[62] As an alternative to these approaches, Weber urged historians to become cultural sociologists, to focus on the interaction of all the three, but especially the civilizational and cultural sphere. This approach allowed for a more open-ended view of history, one that he, unlike his brother and his colleague Mannheim, qualified with a monadic organic premise.

He began a grand sweep of a version of such a history in 1931, and he finished it in late 1934, after the Nazi seizure of power. It was published in the Netherlands in 1935 under the title *Cultural History as Cultural Sociology*. An expanded edition plus three more books elaborating on its themes would follow during the next eight years.[63] Matti Luoma is correct that Weber's descriptions of the vast sweep of "cultural physionomies" in these works never matched the methodological specificity of his earlier article on Egypt and Babylonia.[64] This book can be viewed as an alternative metahistory to Spengler's. It differed in a number of ways, but two stand out. It was more akin to Max Weber's approach in that it emphasized the uniqueness of Western civilization through its dynamism. And it was not pessimistically deterministic. The die for the future had not yet been cast.

Weber dedicated his first chapter of the book to primitive humanity, which he divided into two types: the prehistoric "first person," who was embedded in the natural world, and the primitive, "second person," who began to respond to nature more systematically and to connect himself/herself to transcendent magical forces. From this magical causality "all mythical, metaphysical, and truly religious thinking arose."[65]

Geographical changes and the resulting movement of populations (which Weber might have assigned to the social process, but did not, explicitly) led to the emergence of historical high cultures. He identified the four primary high cultures as the Egyptian, Babylonian, Chinese, and Indian. These witnessed a greater control over nature, the emergence of world religions, and a truly historical humanity. They represented the appearance of the "third man."[66]

To the primary high cultures Weber added secondary ones that emerged with the incursion of new peoples into the old circles. These cultures did not develop out of the primary cultures but established themselves in two great strata on top of them. With their new civilizational, social, and cultural forms, they represented something completely new. The first stratum of these secondary cultures occurred in the Near Eastern and the Mediterranean areas. A second stratum, consisting of cultures that Weber classified grouped in either the "East" (*Morgenland*) or the "West" (*Abendland*).[67]

After 1500, the modern period was examined only for the West, which began to expand throughout the world. The modern, which actually began more than two centuries earlier with the Italian Renaissance, "crystallized" after 1600 in the forms of the modern state, capitalism, and modern science, the three engines of the civilizational process. With this occurrence, nations became "bodies" within their own right and quantitative individualism made its appearance. Weber believed this period represented the culmination of "third person," and reached its own height with the "German Renaissance" of the late eighteenth centuries, the generation of Goethe and Schiller.[68] Two points should be made about this period of renaissance. First, Weber highlighted it in his first publication about culture in 1909, not long after arriving in Heidelberg, holding it up as the model for individual "self-fashioning" (see chapter 4). Second, and connected to the first, the late eighteenth century in Germany was the period that gave rise to the vitalism, the ideal of cultivation, and monadic organicism that characterized Weber's world view.

He wrote that as the nineteenth century progressed, the civilizational process became dominant. This was essentially the rationalization process that his generation, especially his brother, portrayed, and it would descend into the chaos of World War I. Weber's answer to his original question about where they stood in history was: in a time of cultural crisis. He concluded by wondering if this crisis could be resolved, if a new "historical body" based on a "new constellation" of a "new life-aggregation" and a "new culture" could break out of the chaos, or if, as he expressed it in the second edition: "will the fourth person appear?" The fourth person is a fragmented being incapable of cultural synthesis and without freedom.[69] In 1934, he had every right to wonder.

During the republic, Weber wrote not only about his hopes for cultural unity and then the coming cultural crisis, but also on the application of culture to the political scene. The political writings followed the same trajectory as those on culture, beginning with optimism and ending with a much more tenuous perspective.

8
Cultural Politics in Weimar

During the empire, Alfred Weber, like many of his generation, advocated strengthening the lower house of parliament (the *Reichstag*) at the expense of the discursive coalition. He rebutted Schmoller's charge that political parties were merely interest groups in contrast to an interest-free officialdom. He displaced the organic unity of the nation, its "spirit," from the authoritarian state and located it in a more ambiguously defined cultural sphere. With the advent of the republic after World War I, he did not change his stance on the bureaucracy, but, like his brother Max, began to have doubts about the succeeding parliamentary system. By the end of the Weimar Republic he was urging the same limitation of the power of the *Reichstag* that he had attacked upon his arrival in Heidelberg, but without a return to the imperial discursive coalition. In short, by the early 1930s, Weber assigned the state, in both its authoritarian bureaucratic and democratic parliamentary forms, to the civilizational side of his dichotomy, denying that it could provide the necessary organic cultural solutions to Germany's problems. His early hopes that trade unions could separate themselves from party influences and achieve some kind of organic role also dissipated during the republic. The beginnings of this change in perspective can be detected in the last year of the war.

With the political crises of 1917 and questions about the form the German polity would take after the war, Alfred Weber, while not abandoning his discussion of Germany's place in the world, again addressed domestic issues. His brother Max also wrote on the same set of problems. Although there were many similarities in the brothers' positions on what form a postwar government would take, there was a very profound difference between them: Max did not elevate culture to the role Alfred prescribed for it. Instead, he advocated a primarily political position.

The Weber brothers and the question of postwar leadership

In 1918, Max published a long essay, "Parliament and Government in a Reconstructed Germany," that was the culmination of a series of shorter pieces written in 1917. In this work, he applied concepts developed earlier in his career to the situation facing Germany. In January 1919, after the revolution, he delivered his famous "Politics as a Vocation" lecture, which was paired for publication with the earlier "Science as a Vocation." Together, these essays offered a prescription that privileged politics over culture in addressing Germany's future.[1]

As one would expect, Max attacked the discursive coalition that placed the state as an organic whole above civil society. Like other members of his generation, he emphasized the bureaucratic character of the state, and he used the same "steel casing" ("iron cage") metaphor to argue that there wasn't much difference between private capitalist management and bureaucratic management.[2] Although he did not use Alfred's terminology, he clearly located the state within the civilizational process. A passage in his discussion contains language similar to that of Alfred's vitalism:

> An inanimate machine is spirit objectified.[3] Only this provides it with the power to force men into its service and to dominate their everyday working life as completely as is actually the case in the factory. Objectified spirit is also that animated machine, the bureaucratic organization, with its specialization of trained skills, its division of jurisdiction, its rules and hierarchical relations of authority. Together with the inanimate machine it is busy fabricating the casing of bondage which men will perhaps be forced to inhabit one day, as powerless as the fellahs of ancient Egypt.... This casing, which our unsuspecting literati [for example, Schmoller] praise so much, might perhaps be reinforced by fettering every individual to his job..., to his class..., and maybe to his vocation.[4]

Earlier, Max had briefly introduced the concept of the "nation," which he described as a unity anchored in the perceived superiority of its cultural values, and he noted that intellectuals—who possess cultural capital—often designate the nation as a "cultural community" (*Kulturgemeinschaft*).[5] However, in addressing the German crisis, Max, in contrast to Alfred, largely ignored this category. For him the issue was political, which meant the configuration of the state and the conduct of political actors. In contrast to Alfred, Max looked to his brother's civilizational sphere for the solution to Germany's problems.

Like Alfred, Max defended parliamentary government against Schmoller's "constitutional" (authoritarian bureaucratic) establishment. He based his argument on a series of dualistic types. First, he distinguished between politics (making decisions) and administration (implementing decisions). The latter was the task of the bureaucracy, a necessary component of any modern state, and the former belonged to political actors. Decision making in the post-Bismarckean imperial government had fallen to the bureaucracy by default, which proved to be disastrous. This occurred because of the "negative" parliamentary structure of the empire. The government—the cabinet and bureaucracy—was not answerable to parliament, giving the *Reichstag* no role in making decisions. Instead, it simply said "yes" or "no" to them in the name of its various constituencies. Bearing no actual responsibility for constituting a government and devising its policies, members of the *Reichstag* were free to hold to a hard ideological line without compromise. Hence, they never learned the art of governing.

This critique coincided with the emphasis of Max's sociology on individual action and its orientation. In both of his essays of this period, he distinguished those individuals who lived "from" and those who lived "for" politics, acknowledging that these were not mutually exclusive types. Those who lived from politics viewed it as a means to acquire power and income by dispensing patronage. The political parties they formed were designed for this purpose. Max offered the American political "boss" as an example of this type. Those who lived for politics sought to achieve substantial goals in accordance with their world view. These politicians often earned their livelihoods from their political activities, but that was not their primary concern. Their political careers were a means, not an end. These latter type of politicians were again divided into subtypes: those who acted according to an ethic of "conviction" (*Gesinnung*) and those who acted according to an ethic of "responsibility." The system Max proposed was designed to foster the responsible politician.[6]

This politician would be created through a strong parliamentary system. In the give and take of party struggles for power in parliament, the most able leaders would rise to the top. In contrast to the parliamentarians of the empire, who could refuse to compromise their convictions because they had no governmental responsibility, these leaders would have the responsibility of forming a government and of supervising administrators. These demands called for adjusting their convictions to reality.

However, the ideal leader would not be simply the leader of a parliamentary party. Max, following Robert Michels, wrote that parties themselves could become bureaucratized. This was especially the case under a negative parliamentary system. He argued that in a democracy, plebiscitary leaders could emerge, political entrepreneurs whose vision was matched by their ability to win the confidence of the masses. When faced with great tasks, democratic mass parties would submit to such leaders. Parliament could still act as a check on abuses by such a caesarist leader, but political initiative would come from him.[7]

In sum, Max believed that the rigidity of bureaucracy had to be controlled by parliamentary institutions, which, in turn, had to follow the leadership of individuals with a passion for politics but also a grasp on what reality allowed them to accomplish. As with his sociology, he promoted individual behavior rather than organic unity as the key component and politics rather than culture as the sphere of this action. In "Parliament and Government," intellectuals were largely scorned as the "literati," and "Science as a Vocation" was dedicated to setting limits on the role of science in the determination of values. This emphasis, despite some common demands and language, put him on a different track than Alfred.

Alfred's call for a new German spiritual leadership of *Mitteleuropa* in his 1915 book of letters was transferred to the domestic sphere in 1918, when he wrote that with democratization would come the development of both individuality and communality, a new relationship between the leaders and their followers.[8] This was the same theme that he had discussed a year earlier, but now, instead of the leader being Germany and the followers being the smaller nationalities of *Mitteleuropa*, the leaders were the German elites and the followers were the German masses. This focus on the relationship of leaders to followers was, as we have seen, also central to Max's concerns. The difference was in the way they defined leadership.[9]

Like Max, Alfred lamented the "steel casing" of German political and economic institutions. In this essay, he did not use that specific term, but instead used similar ones such as "husk" and "crust."[10] As he did in his earlier cultural writings, he relied on his binary construction and described these institutions as civilizational. But he did not substantially address the issue that had concerned him most before the war and that Max had addressed in his long essay—the bureaucratic state. Nor did he specifically discuss the workings of parties and parliamentary government as Max had done. Instead, he focused on two things—the weaknesses of Western democracy (that of Britain and the United States) and the cultural sphere, especially its leadership. Instead of plebiscitary leaders, he promoted spiritual leaders.

Alfred faulted Western democracy for its civilizational character, which he described as rationalized, mechanized, and atomized. This democracy failed to make a distinction between leaders and the masses. As simply a "voting mechanism with majority results," it was easily manipulated by shallow civilizational forces, such as the capitalist press, democratic cliques, and money in general. It had replaced the old authoritarian system, but offered nothing substantial in its place. In some ways, Western democracy was the opposite of the German authoritarian state: it eliminated the distance between leaders and led, while the latter exaggerated that distance. But in other ways they shared features, for both were described as rationalized mechanisms. This similarity demonstrates Alfred's belief that many of these civilizational elements of government were inescapable; they were simply part of modernity. The issue was how to limit their impact, something that had not happened in the West. The problem in Germany was that those who defended the bureaucratic state simply assumed that all democracy was identical with Western democracy and remained staunchly "antidemocratic." Instead, Alfred advocated a "postdemocratic" solution, one that lay in the sphere of culture.[11]

Alfred did not simply side with the existing cultural, or "spiritual," elite, who, he felt, were part of the problem. He again attacked the old Social Policy of the discursive coalition, this time indirectly. These intellectuals only conducted piecemeal investigations of "problems," such as the "social problem," with a narrow factual radius. Their perspectives were necessarily partial and practical, an approach that conformed to the demands of the civilizational sphere and turned universities into bureaucracies of specialists. While this restriction fostered individual careers, it divorced them from a larger public life. The result was a splintering of both public and political life without an appreciation of the nation as an organic whole.[12] They simply assumed the existence of a larger organic whole located in the state, while engaging in conduct that prevented the realization of such a whole.

Alfred's postdemocratic whole would be embodied in an organic "national faith" (*Volksglaube*). Its leaders had to "penetrate" to a new psychic (*seelisch*) center that would serve as firmer ground and then build from it. This was to be done in conjunction with political democracy. This construction involved separating the part of democratic thought that stems from human self-consciousness—the psychic element—from its rational apparatus. Then, one had to bring out the primal facts of life and combine them with democratic feeling. This effort necessitated a direct intuition (*Anschauung*) of these authentic forces as composing an organic living totality of life. Using the term he would shortly elevate to primacy in his cultural-sociological writings, Alfred described this center as a "new soul" and the totality as "the free inner movement in a psychic unity accessible to all." This soul formed the essence and destiny of the nation.[13]

Alfred believed that the spiritual unity of the nation had begun to emerge with the war and that the potential was present to break through the old system. Its success would depend on the emergence of a new generation of spiritual leaders. These leaders would not simply dictate to the masses as the old leaders had, for the democratic freedom of all the German people was essential to the new configuration. Instead, they had to capture the consciousness and faith of the masses that comprised the soul of the nation. He ended his essay with the hope that such leaders could be found.[14] Three days before a republic was declared on November 9, 1918, he responded to an appeal by the last imperial government under Prince Max of Baden with a call for the dismantling of the old system in order to prevent the "fantasy" of the people from being unleashed with disastrous consequences. Once again, he proclaimed the need for a new spiritual leadership to direct the "stream" of democratic sentiment along constructive channels.[15]

Both Max and Alfred advocated a "leader democracy" (*Führerdemokratie*) that postulated an emotional bond between the leader and his mass followers. However, they differed on what this relationship entailed and on what basis the leader acted. For Max, the leader acted on his own convictions tempered by the constraints of reality. He was not the simple representative of the desires of the masses, but rather the electoral manipulator of them. Whether he was directly elected by the people (as was the president of the republic), which he strongly urged, or indirectly chosen through the parliamentary process, he was above all, responsible to his own individual judgment. This individual responsibility was considerably more muted in Alfred's writings. He assumed the existence of an organic spiritual community along the lines of the proclaimed wartime "national community" and "Ideas of 1914" that would serve as a mediating force between leaders and followers. Whereas Max's political individual leader was someone who stood his ground in an atmosphere of warring political gods, Alfred's spiritual leader was in tune with some larger organic totality.

Alfred offered few details about that totality, but it is clear that his premise was the same monadic organic one that characterized this whole period. In his earlier speech on "Religion and Culture," he had also called for a new cultural faith, but without the political emphasis that is found here. As the war approached its end, Weber looked toward a new democratic Germany and the establishment of a common cultural identity to overcome the civilizational fragmentation that the old regime shared with its Western enemies. He was to be sorely disappointed.

Alfred Weber and the crisis of cultural politics

Weber's prewar writings on the state and Social Policy represented a challenge to the discursive coalition of imperial Germany. With the defeat of Germany in 1918, that coalition no longer existed, so Weber sought to provide guidelines for the new democratic republic. His most important questions involved the composition of the spiritual elite and its relationship to the new political forces, especially the parliamentary parties.

Having designated the most prominent institutional forms of politics as civilizational, Weber placed his faith in "leader democracy" (*Führerdemokratie*), in which an organic relationship between the cultural elite and the larger masses was to be established. For the new postwar Germany, Weber sought a cultural theory of democracy, termed "inegalitarian democracy" or "leader democracy," which would establish a synthetic relationship between leaders and followers.[16]

In doing so, he expanded his brief prewar statement into a book-length theory of the modern state, *The Crisis of the Modern Idea of the State in Europe*. As he had done earlier, Weber assigned the modern state, with its bureaucratic "apparatus," to the civilizational side of his dualism. However, now he made the connection of the state to modern capitalism even stronger than before, with the state becoming mainly an instrument for the development of capitalism.[17] In a passage that echoes Marx, he wrote:

> The modern state and capitalism north of the Alps are actually correlative phenomena, the state being the nourisher and instructor

of capitalism in its invasive formation of new lands. And the state was itself created in its structure, essence, and content first of all by this task.[18]

As part of its symbiotic relationship to capitalism, the state had to define itself in relation to its inhabitants and to other states. The new civilizational aggregation demanded a new "cultural aspiration" to replace the eclipsed traditional legitimism; this was the task of the cultural sphere. In his discussion of the conceptual basis of the state, Weber circuitously addressed the approaches to the issue discussed above in chapter 6 without citing any of the major theorists but instead focusing on the attempts of spiritual elites at the end of the eighteenth century and into the nineteenth century to establish such a cultural aspiration. His discussion includes the major Western polities as well as Germany, but it can be read as an account of the latter's development. Although he presents a number of different principles, these can be condensed into two primary ones. The first is the liberalism of the Enlightenment with its concepts of human rights and constitutional government. Its German form emphasized the separation of the state from society, that is, the constitutional state of law (*Rechtsstaat*). The second, characteristic of the concurrent schools of Romanticism and Idealism, emphasized the idea of nationality, which presupposed the organic unity of the people (*Volk*). Both principles informed the revolutions of 1848, which were also influenced by Western ideas.[19]

Weber does not concern himself greatly with the conservative form these two principles took on during the empire. (We can view them as roughly in accord with Laband's positivism and Gierke's organicism respectively, both supportive of the authoritarian state.) Instead, he discussed the inability to provide orientation in the changing civilizational process. As capitalism spread to the world at large and competition among European nations greatly increased, a new form, monopoly capitalism, emerged with two economic strategies, imperialism and neomercantilism. Without these strategies, Weber claimed, his theory of the location of industries showed that these centers of capitalism would have developed outside Europe. The new monopoly capitalism prevented this spread from occurring through the application of new industrial technology to its military establishments. The result was the growth of the military industrial complex as well as a more chauvinistic racial nationalism. On the eve of World War I, this triumph of the "Power State" (*Machtstaat*), legitimized by a biologized theory of struggle, destroyed the balance of power of the earlier period and gave way to the alliance system.[20]

Not only did the relationship of nations to one another change with the emergence of monopoly capitalism, but the older notable politics within nations also disappeared and, with it, the unity that had characterized the public sphere. Its place was taken by purely mechanized aggregations of individuals who were organized as the masses through a new press and mass political parties. The "spiritual people," that is, the formulators of the earlier concepts of the state, were now cut off from any influence and became disoriented with the revaluation of the old foundational ethical values.[21]

Weber believed that a new spiritual leadership would have to be established to provide cultural meaning for this new civilizational configuration. Despite the similarities in his description of the relationship of the state to capitalism to that of Marxism, he located the solution in the sphere of culture. He applied his monadic organic theory of culture to two broad areas of policy—foreign policy and the domestic policy of democratization. During the war, he emphasized foreign policy over domestic policy, but as the war drew to an end and the prospects for democracy increased, the emphasis was reversed.

Weber did not believe that the arrival of the republic in 1918 automatically meant the realization of his cultural hopes. It was true that the authoritarian state forms had been discredited by the war. Nevertheless, the new "economic state" was unable to create an organic unity out of the modern forces. Liberal notable politics, with its ideal of individualism, proved incapable of incorporating such new forms of mass democratic society as the mass press, party organizations, and the new public opinion. The consequence was that spirit had become separated from politics, leaving politics without the necessary spiritual direction and, therefore, incapable of movement.[22] However, the basis of this new disintegrated public, the new freedom of the masses, could not be removed. Nor did Weber wish for such a thing. He accepted the new democratic civilizational forms, but he also sought a new cultural synthesis, without which he believed the crisis would continue.

Weber implicitly rejected the polar concepts of the state of both Kelsen and Schmitt for the middle formulation that Sinzheimer, Heller, Smend, and Radbruch shared: a belief that the pluralism of republican civil society as embodied in parliamentary democracy could be reconciled with an organic cultural unity of shared values.

In so doing, he looked to a spiritual aristocracy for the medium of this reconciliation. Such an aristocracy was not incompatible with modern democracy. Leaders would be selected by democratic means, but they would be people of such quality that they could put aside conflicting material interests, and would work for the common good. He was not very precise about the relationship of the spiritual leaders to the political leadership in what he termed "leader democracy."[23] At some points, they almost seem identical; in other places, the political leaders would seem to be informed by the spiritual leaders. It appears that the two groups would participate in both the cultural and civilizational spheres. Because they were part of the same cultural context, which was superior to the civilizational process, the spiritual leaders could give direction to the political leadership within that sphere, and this meaning could then give direction to the institutions of the civilizational sphere.

Weber's cultural analysis contained elements from both types of public spheres. Like the bureaucratic type, it assumed an organic synthesis that provided a direction for the civil society. However, like the liberal type, the public sphere had primacy over the state, which was parliamentary, in the determination of cultural meaning. He also blurred the relationship of intellectuals to political actors. Such a position was consistent with his cultural sociological thesis that culture formed an organic whole and gave meaning to the historical body. Just as the cultural sociologist could separate the two spheres in his analysis, so the leadership, which participated in both the cultural and civilizational spheres, could separate those activities, and achieve a cultural synthesis. This would be possible, he wrote, only if:

> there is a fluidity, or if a fluidity can be created, which, in spite of the disruptive influence of the materially and partially limited forces streaming through the body of the state, still bears it as a unity that moves back and forth above the interest groups and that forms a general something in all that is essential to the life of the whole. The political leader must come forward as the exponent of this fluidity, who wants to take hold of it and who, therefore, actually holds it together and activates it in political action.[24]

Weber believed that this necessary spiritual leadership would emerge from the youthful elements of the various parties and assume a spiritual-aristocratic posture. The latter could separate personal standards from party opinion and could work out a spiritual-aristocratic norm.[25] He hoped that intellectuals, in their role as the formulators of the cultural totality, would provide a solution to the cultural crisis. But to do this, they would have to overcome their own existential crisis.

Weber, Bourdieu, and the crisis of the intellectuals

In late September 1922, Weber examined this crisis in the keynote address, titled "The Distress of the Spiritual Workers," at the annual meeting of the Social Policy Association. The country had recently moved from significant inflation to hyperinflation with the mark-to-dollar exchange rate almost tripling, and Weber examined the impact of this phenomenon on intellectuals. Bernd Widdig, in an excellent analysis of this talk comparing it to the cultural sociology of Pierre Bourdieu,[26] notes that Weber's was the most widely discussed treatment of the inflation dilemma. Examination of the content of Weber's lecture and then of Widdig's treatment of it is useful for two reasons: the talk is Weber's most specific application of his cultural sociology to the concrete reality of his time, and the comparison to Bourdieu addresses the potential significance of his work.

In his study, accompanied by a 14-page statistical appendix, Weber described the vocational and material toll that the inflation took, on spiritual workers. During the empire, members of this group were primarily "rentier intellectuals," meaning that their income was largely independent of immediate economic circumstances. They came from the upper strata of the bourgeoisie that had amassed significant capital, or they were high state officials or academics who (as noted earlier) had very dependable and comfortable incomes, which allowed them to focus on spiritual issues. In addition to security and independence, their position also conferred social and cultural status. This situation changed with the economic conditions. Because their wealth was relatively fixed, it was severely devalued by the inflation. Family fortunes were wiped out. Those on salary saw their incomes go from seven times that of a manual worker to almost equal to that of the worker. Weber argued that this economic leveling contributed to a leveling of status.[27]

Weber wrote that these intellectuals, "the proper stratum of cultivation and ideas in the stricter sense," were embedded in a larger public, made up of the cultivated elites (*Bildungsbürgertum*), which in turn was intertwined with the social order. He emphasized that this larger cultivated public did not form an encapsulated entity at the top of the social pyramid, but rather was "something blurred and opaquely interwoven, grown together with other strata." The larger cultivated public also suffered hard times in the inflation, and, as a result, its demand for the cultural products of the spiritual elite was reduced. Academic publishing venues decreased from lack of demand as did the market for high art and literature. Equally important, the cultural infrastructure upon which spiritual workers depended—libraries, institutes, research facilities, museums, and the like—lost a considerable amount of their funding.[28]

A larger mass audience that was less discerning took the place of this cultivated public, and the cultural creations of the spiritual elites were forced to compete with those aimed at this larger audience. The resulting decline in "purely spiritual literature" was accompanied by an increase of "spiritual narcotics," such as mystical and superficial literature. The increased influence of industrial magnates in the cultural sphere accompanied this process. Intellectuals were forced to cater to the wishes of these "industrial princes," further demeaning their creative abilities and casting doubt on the value of pure intellectual products that had not become commodities. The result was a loss of intellectual freedom, which further deprived intellectuals of their status and influence.[29]

Weber believed that this change in publics doomed the old type of intellectual. But he held out the hope that an alternative to it, the "worker intellectual," would appear. This type would ascend from and address itself to the growing stratum of white-collar workers. It would combine a spiritual orientation with a more practical vocation. These individuals would resemble and include lawyers, doctors, and engineers in that they were essential to modern society and would also contain a strong spiritual component. Weber suggested that these intellectuals could unionize, while admitting that there would be limits to this, given the individual nature of cultural creation. He believed that the establishment of a minimum fee would still allow for individual freedom. He indicated that higher officials such as professors, the very core of the spiritual elite, could probably not make this transition, but that they would be willing to sacrifice some of their earlier comfort out of a sense of duty to the whole. He also noted by way of a hint that only a small percentage of state budgets were consumed by their salaries, so state governments could afford raises. He believed that this new type of spiritual worker also would be able to grasp and formulate the "public spirit," which "today no longer sits in the official chambers where fifty years ago it was sought out by the Social Policy Association in order to influence it." Instead, it lived in the organs of the democratic state, in the press and the parties.[30]

Widdig argues that both Weber and Bourdieu investigated the relationship between cultural production and social and economic power. Bourdieu distinguished between "restricted [cultural] production"—"production for producers"—and "large-scale production" aimed at a mass audience. The latter generates revenues, that is, economic capital, while the former creates "symbolic profit," or "cultural capital." Each of these fields has its own institutions: museums, libraries, and universities in one case and movies, lifestyle magazines, and bestseller publishing in the other. Money, the most fluid form of economic capital, plays a determining role in the latter but not in the former, where there is considerably more freedom ("art for art's sake"). Nevertheless, this freedom translates into a distinction that delivers cultural capital and symbolic power. Bourdieu accordingly views the possessors of this power as part of the dominant class.[31] Widdig concludes:

> It is remarkable how Alfred Weber's speech of 1922 unintentionally employs Bourdieu's theoretical parameter.... [Weber] identifies the dominant class as the main consumers of intellectual work and the main providers for the next generation of "intellectual workers."... Weber shares with Bourdieu the same understanding of the paradoxical position of critical high culture within capitalism.[32]

Insightful as this comparison is, it leaves out an important contrast between the two men. Bourdieu's main concern is with power, specifically the power held by an individual within a "field." This power involves interactions with actors outside the specific field. For example, he writes that an alliance with the possessors of significant economic capital could increase the symbolic power of some cultural producers over other cultural producers. Conversely, a limited market for one's production due to limited demand could actually increase one's prestige and symbolic power.[33] These possibilities occur either within the field in which the individual is situated or between that field and other fields. This interaction amounts to a set of conflict relationships between these individuals. Bourdieu's conflict model is essentially class-based, despite the fact that class and power are defined in a broader manner than Marx's economic criteria so that the relationship of the economic and cultural spheres is not reduced to simple determinism. It is about competition for power between individuals with differing access to various forms of capital.

Weber's model is, conversely, an organic one in which conflict is overcome through cultural production. Spiritual unity, rather than distinct forms of individual or group power, is at the heart of his approach. The symbolic power acquired by individual intellectuals was not an end in itself, as it was for Bourdieu, but a means to the end of that organic unity. Both men portray two qualitatively different spheres, the cultural and the economic-civilizational, which interact with one another. For Bourdieu the two spheres are defined by a competition for power, while for Weber only the latter is. That sphere would ideally be transcended in an organic cultural unity. Accordingly, Weber believed that the new rentier intelligentsia, "despite all its inner contradictions, still on the whole formed a type of spiritual unity." Although these individuals had material connections to the fragmented civil society, their interaction with one another could establish an organic meaningful unity for their time. His hope was that the new worker intelligentsia would be able to preserve this unity, which was essential even for the viability of the civilizational state.[34] The ability of the cultural elite to transcend the divisions of society makes his intelligentsia socially free-floating in the manner often incorrectly attributed to the well-known concept of his student Karl Mannheim.

Weber and Mannheim

In 1921, Mannheim voluntarily left his native Hungary after the failure of a Communist revolution there and 1922 settled in Heidelberg, where he was sponsored as a lecturer by Alfred Weber and Emil Lederer. An economist who was active in socialist circles, Lederer, like Mannheim, was a Jewish emigré from the old Hapsburg Empire, where his mentors were members of the Austrian School of Economics. He became a coeditor of Max Weber's *Archiv* after the latter's death, and he was also codirector, with Alfred Weber, of the InSoSta. Both he and Afred Weber supervised Mannheim's habilitation thesis on conservatism.

During Mannheim's first years in Heidelberg, he was a member of Alfred Weber's cultural sociological seminar. His connection to Lukács gave him a certain cultural capital, which he compounded with his intellect and a series of impressive publications, and he attracted a number of students, including Norbert Elias and Hans Gerth. Early on, he largely followed Weber's postulation of a unified organic cultural realm. In a long unpublished manuscript of 1924 that was written in the seminar, he closely adhered to Alfred Weber's view of culture: he distinguished between two types of knowledge, "communicative" (which was civilizational in that it was supracultural and unilinear) and "conjunctive" (which was cultural in that it was limited to a given historical context). His view of intellectuals, presented in the concept of "cultivated culture" (*Bildungskultur*), was also similar to Weber's. The cultivated culture occupied a position between the purely objective civilized sphere and "the existential rootedness of cultural creation." The cultivated culture allowed for a certain distance from the cultural "life-community," taking the germs (*Keime*) of the latter to a level where they interacted with others. The cultivated elite who formed the cultivated culture were relatively independent from their original primary communities and were able to interact with one another spiritually, which allowed for a "widening of horizons" and "heightened sensibility." But this higher level was not "free-floating" (*freischwebend*), because it combined cultural tendencies within the same historical continuum. Nor could it "generate new germs, a new stress, on its own." It allowed for an increased consciousness of the place of its participants and their original communities within the stream of history, but "while it transcends community in one sense, cultivated culture remains conjunctive nevertheless."[35]

Although one can see the beginnings of his later theory of the socially free-floating intelligentsia (with which Weber disagreed, despite Mannheim's attribution of it to him), the short and nonpolitical concept of cultivated culture remained within the scope of Weber's work. It was based on Weber's culture-civilization dichotomy; it used the same type of vitalistic language Weber used; and, most importantly, it maintained the premise of an organic synthesis. The manuscript reads as a more detailed examination of Weber's cultural sociology.

However, the same year that it was written, Mannheim began a series of publications, culminating in *Ideology and Utopia* five years later, which challenged Weber's organic premise. During this period, he introduced his sociology of knowledge as an alternative to the sociology of culture and gave it a political dimension missing in his earlier work.[36] It was the latter that would bring the two men into conflict. Nevertheless, Weber tolerated this gradual movement away from his position and cosponsored Mannheim's habilitation thesis. Mannheim might have been encouraged in this attempt to negotiate this path between Marxism and the sociology of Alfred Weber by Lederer's earlier essay on cultural sociology.

Lederer argued that the primary task of cultural sociology is to establish the relationship between cultural objectifications—whether it involves a work of art or a legal concept—and the social. Rather than making pronouncements on the "essence" of a work, the sociologist should clarify the presuppositions of its "time" (its specific historical sociological structure) and the conditions for its realization. This approach mandated moving away from the methodologies and formulations of specialized fields, such as aesthetic judgments or legal formalism, to examine how art or law (and artists and jurists) are "embedded" in the structure of economic and social life.[37]

To accomplish this, the sociologist must identify the totality of social relations that comprised the time. Lederer identified two approaches to the latter: the Marxist, which defined a time strictly in terms of economic relationships so that people were simply the bearers of economic interests, and the non–Marxist, such as that of Alfred Weber, which included more than just economic relationships of production in the totality. He joined the Weber brothers in criticizing the reductive nature of Marxist explanation, which saw the cultural realm as simply a reflective superstructure; and he wanted to redefine the social to include the spiritual as well as the economic, arguing that it can also be included in the category of productive relationships. Anticipating Pierre Bourdieu, he wrote that the term "capital" can take multiple forms that include the qualitative as well as the quantitative. Such a reformulation of the social would mean that epochs would be defined as shorter periods of time, since these more complex interactions changed more rapidly than did the economic structure.[38]

Lederer argued that although the social sphere does form the foundation for culture, it cannot be viewed as a "totality." Instead, it is a very complex set of interactions that constantly change. It is useless to view it as a set of regularities. In some cases, there is a direct correlation of cultural forms with social strata, but not always. For example, the social base for expressionism existed in the late nineteenth century, but it was only with the fragmentation of the war and the revolution that it came to the fore. Furthermore, one cultural object can be enclosed in another cultural field—for example, art in religion—weakening the impact of the social base. Finally, the socioeconomic base can change in such a way as to lessen its impact on culture. As an example he argued that the increasing abstractness of relationships in modern capitalism weakens the bond between the social base and cultural creation, giving the spiritual sphere a freedom and lack of direction, and, with it, the possibility of disorientation.[39] Lederer's unwillingness to portray culture as an organic totality was echoed in Mannheim's writings.

The divergence of Mannheim from Weber became public in 1928 at the Sixth Congress of German Sociologists, where Mannheim delivered a paper, "Competition as a Cultural Phenomenon." He followed Weber by dividing thought into two spheres, that of the exact natural sciences (which might be termed civilizational) and that which was "existentially connected," that is, cultural.[40] Like Weber, he defined the cultural sphere as the realm where meaning and values are located. However, he depicted this sphere as a "polarized" competition, by which he meant that different social elements were organized into parties, each subscribing to a "spiritual current," or world view, and striving to determine the "public interpretation of existence." He vaguely identified three such polarized parties as conservatism, liberalism, and socialism, which were defined not by economic interests or organizational structure, but by their commitment to "spiritual currents."[41]

In the discussion that followed, Weber leveled his criticism.

> What I find missing in your discussion is a recognition of spiritual creativity as a basis for action, for example for the action of social classes. What I reject is the reduction of all these things to, in the final analysis, intellectual categories to which have been added several—if you will excuse me—sociological categories taken from the old materialistic view of history.... Is all this anything more than a brilliant rendition of the old historical materialism, presented with extraordinary subtlety? Basically it is nothing else.[42]

Despite the phrasing of this passage, Weber attacked Mannheim not for being an economic determinist, but for practicing a form of "intellectualism," the division of an organic unity into analytical categories. Mannheim, like Marx, was postulating a pluralistic conflict rather than any kind of unity in the cultural sphere, and it was this fragmentation that drew Weber's ire. Instead, he advocated "spiritual creativity," which meant the ability to create a cultural synthesis of meaning, precisely what Mannheim rejected.[43]

This debate continued early the next year in a joint seminar the two held in Heidelberg on Lukács's *History and Class Consciousness*. Weber repeated his charge of intellectualism, the analytical reduction of cultural unity into divisive categories akin to Marx's historical analysis. Such categories, said Weber, whether portrayed as material interests or as interpretations of existence, mean the fragmentation of the larger organic sphere of meaning. Acknowledging his vitalist premises, Weber claimed that only spiritual creativity could bridge the gap between that organic whole and the soul of the individual.[44]

Mannheim responded to Weber's depiction of his approach as intellectualistic by stating that he and Weber were presenting different conceptions of rationality. His approach was "functionalist," and was oriented toward the more modern model of "achievement" in which "the rationally achievable and controllable comes ever more to the fore." Weber, on the other hand, was engaging in a "morphological" attempt to rescue the irrational, which was oriented to the model of "growing," and which attempted to leave a "substantial being" untouched other than nurturing it. Adding a touch of irony, Mannheim declared that in sociology, the methodology representing his approach was that of Max Weber.[45]

Mannheim's pitting Alfred Weber against his brother would continue in *Ideology and Utopia*, which added nothing essentially new to his presentation of the structure of the cultural realm. Rather, the book would work out many of the implications of that structure, including the role of intellectuals in that polarized realm of culture.

Although Mannheim attributed the concept of the socially free-floating intelligentsia to Alfred, he based the concept on Max's rejection of an organic cultural totality. His intellectuals represented a multiplicity of groups and commitments, but their qualities as intellectuals prevented them from entering completely any of the competing sociopolitical groups. "The[ir] repeated attempts to identify themselves with, as well as the continual rebuffs received from, [classes and parties] must lead eventually to a clearer conception on the part of the intellectuals of the meaning and value of their position in the social realm."[46]

The same qualities that deprived intellectuals of full membership in political parties allowed them to interact with one another despite their different social commitments. Mannheim hoped that this shared intellectuality would allow them to understand the interrelationships of their different world aspirations. The intelligentsia was homogeneous and heterogeneous at the same time. The heterogeneity of its members came from individual commitment to the aspirations of their respective groups or parties. The homogeneity came from their common level of cultivation, which allowed the possibility of communication with one another, something not shared with the general population. Thus, the "synthesis" they achieved was a "dynamic" one, in which conflict among commitments was clarified and, to some degree, reconciled, but never transcended in an organic unity. Rather than subordinate political conflict, Mannheim's socially free-floating intelligentsia could only incorporate it.

Because the intelligentsia was free-floating, not bound to any one aspiration, Mannheim believed that it could not be politically committed as a group. Its members could not be the promulgators of political change, which resulted from groups (parties) committed to certain world aspirations. Such groups were effective in bringing about (or resisting) change because all of their members were committed to the same world aspirations. (In fact, this commitment was what gave the group its identity.) But members of the intelligentsia, characterized by heterogeneity, could not be committed to the same political changes. Their common will was not the political will to change the world but the intellectual will to clarify political positions. They were not politicians but scientists of politics.

Political scientists could not actually make political decisions themselves. All they could do was show the context between decision and perspective for the political actors, with whom the actual decisions remained. Unlike Alfred Weber, who, because he believed that the spiritual leadership could rise above pluralistic conflict, blurred the relationship between the elites of culture and political parties, Mannheim, like Max Weber in his vocation essays, sharply delineated the roles of the politician and the scientist, limiting the role of the latter.[47]

Mannheim and Max recognized that the decentered political and economic sphere was accompanied by a decentered cultural sphere. The latter could not form an organic totality, but rather was characterized by conflict, the struggle for cultural hegemony, in which intellectuals played the more restrained role of advisors rather than the creative role postulated by Alfred. Mannheim denied the possibility of a privileged center from which one could interpret this competition. Although Mannheim's intelligentsia could clarify temporary constellations within the competition and provided a medium for communication between the competing groups, they could not grant a privileged position to any of those groups. In short, they could not become "spiritual leaders." This was a position that Alfred Weber could not accept.

Alfred belonged to the generation that challenged the orthodox mandarin discursive coalition by placing the state and officialdom on the civilizational, social side of the organic-mechanistic dualism. However, he retained not only the dualism itself but also the ideal of an organic cultural synthesis. He rejected the old allegedly synthetic institutions, the state and the officialdom, but he was not able to accept completely the pluralism of the Weimar parties. Therefore, he had to seek synthesis in the cultural sphere. Although he recognized the pluralism of the republic, Alfred retained the hope that this pluralism could be centered in culture. This same organic premise continued to organize his thoughts on foreign policy as well.

Foreign policy in Weimar

As they did during the war, Weber's writings on domestic and foreign policy mirrored one another, both being informed by his cultural theories. And since the cultural theory remained largely unchanged, so did the essence of both domestic and foreign policy. Just as Germany was to assume a position of spiritual leadership in Europe, it had to establish it own internal spiritual leadership, which could give direction to the different social groups.[48] During the entire period from before the war through the republic, Weber followed Friedrich Naumann and applied the same pattern to both domestic issues and foreign policy, focusing on the relationship between the leaders (*Führer*) and the rest of the community.[49] What changed—dramatically—were the situations in which the policies would have to be realized. In the larger community of nations, Germany was seen as playing the same role that the cultural elite was to play within Germany. Just as the cultural elite's position had been devalued in the republic, with the loss in the war and the Treaty of Versailles, so had the role of Germany in the world.

In placing Germany at the center of the new European aspiration, Weber charged his country with the same restoration of the balance between the spiritual and the material that he urged on the domestic scene. He carried over the ideals of *Mitteleuropa* into his writings during the Weimar Republic, but the ideals were now applied to Europe itself. The policy of "Eastern Europe" (imperialistic expansion into that region) continued, but now included recovering land in the east lost to new nations such as Poland and Czechoslovakia. Rather than being the dominant policy at this time, as it had been during the war, it was advocated by the far right, especially the conservative nationalists centered in the German National People's Party. World policy, which had led to the war, was replaced in the mid-1920s by the attempt of Gustav Stresemann to tie the German economy to that of the United States, which meant a degree of political reconciliation. Once again, Weber's position fit somewhere between these two.

As he did during the war, Weber placed Germany in a dualistic scenario, but with differences in the dualism. Earlier, he had contrasted the European and Russian cultural spheres and divided the European sphere into the British and German positions. Now, he moved Britain out of Europe proper and designated it as part of the Anglo-American world.[50] Russia was connected to Asia despite the Communist revolution, which he tended to ignore. He located Europe between these two spheres and monadically related it to *Mitteleuropa*, which in turn was monadically related to Germany. Even Germany's diplomatic isolation through its exclusion from the League of Nations favored *Mitteleuropa*, since Weber branded the League as an anti-European institution.[51] He viewed the sanctions against Germany's economy as the deprivation of the economic base of *Mitteleuropa* and, hence, a blow to the viability of Europe.[52]

Weber characterized the Anglo-American world sphere as empiricism, pragmatism, and imperialism—in short, by all of the characteristics that he assigned to the civilizational process. The Russian-Asian sphere represented its antithesis, a spiritual and psychic entity that was transcendent rather than practical, that is, largely cultural. Weber believed that both of these extremes denied the European essence, and for Europe (and Germany) to be forced to choose between East and West meant its demise.[53] Germany, as the most European of the European lands, contained aspects of both the transcendent and pragmatic spheres existing side by side but, at the same time, disconnected from one another.[54]

Because Germany was the only country with both a strong civilizational potential as well as a cultural depth, only it could overcome this dualism and become the cultural center of Europe. The universalistic imperialism of the Anglo-American world might be blind to the cultural variety present in Europe, but Germany, itself characterized by variety, could place itself at the cultural center of Europe by promoting that variety.[55] Such a statement extended the monadic organicism that characterized south German nationalism beyond Germany to Europe as a whole.

Weber believed that the European cause could be furthered by a return to the values of the German revolutions of 1848. The men of 1848, while German nationalists, did not view Europe as a problem. Inspired by the German idealists of the late eighteenth century, they believed in two principles—political multiplicity and spiritual unity, the latter characterized by the belief in general human rights. Weber believed that the balance of power incorporated both principles in which nations sought their own self-realization but respected that of others. The defense of humanity and the nation was the same thing; in the national Idea there was a general mission with supranational meaning. This is why the men of 1848 supported the freedom of Poland as well as of national minorities within Germany. This spiritual harmony had been disrupted with the rise of materialistic imperialism, which Weber identified with the World Policy of both the Anglo-American world and the German Empire. The embodiment of this imperialism in the power state resulted in the separation of the Idea of humanity from that of nationality and biologized the latter into crude racial terms.[56]

Weber wrote that with the decline of imperialism in the larger world, through the emergence of anticolonialist movements, the spiritual will of Europe could once more assert itself over the material-mechanical. As other parts of the world were establishing themselves as historical bodies, Europe had to do the same thing. It had to establish itself as a unity, as something spiritually different and self-contained, with its own "cultural aspiration." Weber believed that the idea of individual human rights remained at the core of this cultural aspiration, which would be realized in some type of federalism. In fighting for the rights it had been denied by the Versailles Treaty and its enforcement, Germany, at the same time, defended European humanity. While this struggle could not be accomplished solely by spiritual means, hence his concern with Germany's economic position, Weber believed that such means provided the necessary cultural core. However, the principles of humanity could not be championed merely in empty political slogans. Rather, "it is a matter of the rediscovery of the direct psychic stream of life from which all these formulations arise and making it again effective."[57]

During the more prosperous middle years of the republic, Weber was optimistic that the new Europe with Germany in the lead would be realized. However, during the depression years and the growth of fascism, he appeared much less convinced that a new European cultural aspiration was just around the corner.[58]

The end of the republic

During the last years of the republic, Weber's optimism that a new cultural synthesis could be created decreased markedly, just as it had in his writings on foreign policy. He published very little on spiritual leadership during this period, indicating that he believed that the written word was not the most effective forum for his position. With the exception of a 1931 encyclopedia article on cultural sociology, most of what was published were speeches and lectures, the longest being a 19-page lecture titled *The End of Democracy?*

This title expressed his concerns. And, in raising this question here and in other pieces from the same period, he indicated a retreat from his criticism of Schmoller 24 years earlier. Then, he had opposed parliamentary government to the authoritarian bureaucratic state. Now, he no longer believed in a parliamentary government based on political parties.[59] A newspaper report of one of his lectures noted that:

> Above all he demanded a reform of our party system.... We have too many parties and *our large parties have no notion of the values of the people.* Weber especially referred to Social Democracy, which to a large part has become a party of functionaries.[60]

Here, one sees parties being described with the same language of "civilization" that he had used to describe the bureaucratic state. Such institutions, he believed, were incapable of the formation of the new type of person that Germany needed. As an alternative, he proposed smaller groups that could come together in a federation (*Bund der Bünde*).[61] He also advocated strategies for taking power away from the *Reichstag*, where the parties participated in the government, in order to preserve democracy, including freeing the executive from parliamentary control by removing the *Reichstag's* power to vote no confidence in the government and by putting the direction of the economy in the hands of the appropriate ministries, whose members were appointed by the President.[62] He denied that this represented a return to the old authoritarian state of the empire, writing:

> Perhaps one will label what is represented in this lecture, which was given two times to democratic students, as "authoritarian democracy." One may do that. The demarcation of limits for that authoritarianism has been given clearly enough. And I conceded equally the necessity for the infusion of backbone and an emphasis on qualities that lead to personal responsibility into the faded and feeble intellectual and actual property of the democratic aspiration, which until now has too much emphasized quantity.[63]

It is ironic that the vigorous antifascist Weber should use terms, like "backbone" and "personal responsibility," that were stock-in-trade for the National Socialists. His usage reveals the tragic aspect of the monadic organicism that he and many others refused to abandon. He admitted that the National Socialists were better equipped to deal with the new reality because they had developed a new vision and a new "*élan.*" What he sought was a new democratic leadership capable of contesting the fascist vision. Such a force could not be purely civilizational as both the old bureaucratic state and the new parliamentary state were. Rather, it had to envision the nation as a cultural community and win over a German people that had lost, not its sense, but its patience.[64] The belief that the early years of the war had witnessed a oneness that could overcome societal divisions, the "civil peace," was proof that such a basic cultural unity existed. As he became frustrated with the pluralistic stalemate of parties, he placed more emphasis on organic solutions, the spiritual leaders, and, ultimately, a more authoritarian political force. Shortly before the July 1932 election, he urged members of the Democratic Party, which had renamed itself the State Party, to cast their votes for the current chancellor, Heinrich Brüning, who had used the emergency provisions of Article 48 of the constitution to bypass parliamentary decision making. This leadership, or any other capable of recharging Weimar democracy, never emerged. Contrary to his hopes, the type of leadership Weber advocated dedicated itself to undermining the republic. The Social Democrats, whom he castigated as party functionaries, were the only ones to stand true. With the demise of the republic, Weber largely retreated from politics. His expulsion from academia would not be long in following. Rather than culture informing politics, the political will perverted the cultural person.

9
Epilogue: Alfred Weber After 1933

From the time he arrived in Heidelberg in 1907, Alfred Weber was fairly consistent in his cultural and political world view, despite the dramatic transformations in the context in which the world view was articulated. Many of the particulars of his position changed with the context, but the overall premise remained the same. Like many of his generation, he accepted the organic-mechanistic dualism that underlay the discursive coalition of the empire, but modified the assignment of institutions to the two spheres. The authoritarian state, which the coalition viewed as the embodiment of the organic unity of the nation, was transferred to the mechanistic sphere, where it now accompanied civil society rather than hierarchically subordinating it. This transfer put the academic elite in a less defined position, for they had been considered a part of the state establishment, even if they were clearly not equals of the main political actors.

Weber departed from those of his generation who adopted the schema of Ferdinand Tönnies and transformed the dualistic model from a hierarchical one to a successional one. These thinkers tended to draw rather pessimistic conclusions from their model, often expressed as a loss of community. Weber attempted to avoid those conclusions by portraying the organic and mechanistic spheres as interacting with one another in a way that was neither hierarchical nor successional. His model elevated the cultural elites in importance and placed them in opposition to the bureaucratic establishment, despite many of them (including Weber himself) being a part of that establishment.

I have described Weber's model as a monadic organic one that was in accord with the setting of Heidelberg as well the tradition of cultivation (*Bildung*), especially in the latter's vitalist forms. This approach allowed him to emphasize both organic unity and free individual creativity without contradiction. At the same time, this model was difficult to translate into any practical political agenda. During the empire, most of Weber's political writings either attacked bureaucratically structured institutions or called for a revitalized cultural elite to serve as a model for the nation. He believed that this elite could constitute an organic center that could coexist with a parliamentary system. Cultural sociology was designed to aid in that task. To a certain extent, he reversed the relation of cultural and political actors to one another from that of the discursive coalition. In foreign policy, Germany was to play the same role for a central European confederation that the cultural elites played in Germany.

The change that resulted from the establishment of the Weimar Republic was not accompanied by any change in Weber's world view. He continued to believe that a cultural elite was compatible with a parliamentary government. He wrote that the elite had to become more inclusive and the parliamentary parties less bureaucratic to achieve success, but monadic organic unity was possible. He elaborated further on the spheres investigated by cultural sociology and the method of examining their interaction, but this did not change the basic premise. In the late republic, as parties became more polarized rather than unified, as many within the cultural elites as a whole became less committed to organic unity than he envisioned, he blamed the parliamentary actors (especially the SPD) for their bureaucratic tendencies. At the same time, his cultural sociology became more metahistorical. That trend in his cultural writings continued after 1933, and especially after 1945.

Weber and the crisis of German sociology

In these cultural writings, Weber hoped to situate the present in the larger flow of history, convinced that such an endeavor would provide orientation for modern humanity and awaken people to the crisis they faced. His position is demonstrated by the opening lines of the conclusion of one of his very last works:

> Sociology is the daughter of crisis, the greatest life-crisis that the West has yet experienced. There is little doubt that at the time of its origin the Western person experienced for the first time what history really is. History, which in its past and future is an immediately essential part of one's present and own self in which self-understanding is like an eye of a needle (Sombart) through which one's destiny runs with history.[1]

This period of crisis had by no means concluded and humanity was faced with the danger that its "hitherto existing psychic-spiritually integrated form" would disappear, to be replaced by nihilism, chaos, and the feeling of abandonment. This quandary was the one that Weber had continually emphasized throughout most of his career: the inability of the cultural sphere to keep abreast of the rapid technical progress in the civilizational sphere, resulting in the loss of its ability to face reality in a creative and formative manner. The task of sociology was to bring about an alignment of the existing specialized disciplines and, through this, to illuminate the problematic in its totality. Sociology, through its cultural and historical approach, could present to humanity its options: it could either remain trapped in an empty civilizational housing epitomized by bureaucratization and nihilism (the fourth man), or it could reinvigorate the forces of life through culture in a way that gave meaning to those civilizational forms (the third man).[2]

Weber believed that the coming of mass democracy (while inevitable and desirable), combined with the dominance of the civilizational process through technical innovation and bureaucratization, had weakened the cultural sphere's ability to provide larger meaning for people. This imbalance had resulted in a lessening of the individual's ability to orient herself/himself in the rapidly transforming world. With the defeat in the war and the end of the Nazi regime, he believed that a modern, revitalized version of the "third man" could be created.[3] The creation of a new democratic citizenry that would be able to understand its place in the world was central to his agenda. He advocated a sociological initiative in education and emphasized the importance of elites in the establishment of a new orientation based on shared values. But Weber never relinquished his belief in vitalistically infused monadic organicism. As a result, the analytical concepts that characterized his brother's work were missing. Klaus von Beyme tells of Weber participating in a 1954 conference on his brother Max:

> There his methodology was sharply attacked by the American sociologist Talcott Parsons [who had studied with him in Heidelberg in the 1920s]. Completely despondent, Weber finally said to him, "do you want to cast doubt on my entire life work?" And Parsons politely responded, "Not at all, dear colleague, but I simply wouldn't call it sociology."[4]

Parsons was certainly correct with regard to Weber's later metasociology, but a case can be made for revisiting the earlier cultural writings, especially if one departs from Weber's insistence that his concepts were not ideal-typical. Both the study of the economic distress of intellectuals during the Weimar inflation and his 1926 study of ancient Egypt provide examples of the potential application of his concepts. The cultural work of Mannheim and Elias departed from Weber's monadic organicism, but they were originally conceived within the sphere of his cultural sociology. The misfortune is that while they continued to evolve, Weber himself ignored the potential of his early work in favor of the metasociology. He abandoned competition with his brother for that with Spengler.

Ironically, his most lasting impact in postwar Germany came from following his brother's earlier call for political actors to live according to the ethic of responsibility. His political contributions outweighed his cultural sociology. His pleas for the establishment of a monadic organic unity went ignored, but his dedication to the new democracy in his actions, while based on his organic beliefs, separated him from many others who wished to simply ignore the preceding 12 years.

Weber and the other Heidelberg myth

After the war, Weber became an active participant in the new Germany at the age of 77. He continued to be part of a number of the town's circles, including Marianne Weber's. He attempted to restart the "Sociological Evenings," but without success. He participated in groups dedicated to democratization, such as the Heidelberg section of the Cultural League for the Democratic Renewal of Germany and the Action Group for Democracy and Free Socialism. From 1945 to 1949, the latter group published a journal, *The Transformation* (*Die Wandlung*), dedicated to democratic issues such as tolerance. He also joined the German Society of Voters, which declined any party affiliation and promoted majoritarian over proportional representation.[5] These actions can be seen as an attempt by Weber to revitalize the old "Spirit of Heidelberg," with its monadic organic premise.

That premise, upon which the "Myth of Heidelberg" had been constructed, also informed many of the personal positions that Weber took, regarding foreign and domestic policy. With a brief divergence after the outbreak of the Korean War, he sought a neutral position for Germany in the Cold War, believing that this was the best way for German unification. This position was the basis for his animosity toward the Adenauer government. He joined the SPD, partly to promote democratic political reconciliation, but disagreed with the party on domestic economic issues. He opposed its policy of "planned economy" and advocated, instead, "free socialism," which was very much in accord with Adenauer's social market capitalism.[6] This position was also located between the Cold War ideological poles of socialism and unrestrained market capitalism. He viewed the possibility of European economic unification as a way beyond nationalism. These positions, which sought a middle way between East and West, were logical successors to his positions in Weimar and, in part, were based on his hope for some kind of organic cultural unity for Germany, for central Europe, and, finally, for Europe as a whole. In general, his positions at the time were unpopular and he had little political influence, even within his own party. Where his actions were most consequential was in the process of denazification, especially of the social sciences at the University of Heidelberg.[7] In so doing, he was forced to confront the other "myth" of Heidelberg.

Steven Remy has written a scathing critique of the university during and after the National Socialist regime. He argues that its faculty very quickly accommodated themselves and the institution to the new political realities of Nazism. Although 28 percent of the faculty, the third highest in Germany, lost their positions, most on the basis of "race," the response of their colleagues to this purge was almost nonexistent. They engaged in no form of collective protest. Many of those who remained joined the NSDAP, either for reasons of ideological sympathy, or pure opportunism, or some mixture of both. And many in this group used their "science" to help legitimize the Nazi agenda. Others joined those who had lost their positions but not their citizenship in "internal emigration." They turned to "safe" topics that they hoped would not catch the authorities' attention.[8]

Remy further argues that, after 1945, there was a largely successful attempt to whitewash this record with the construction of the "Heidelberg Myth," which consisted of an insistence that before 1933, the university had been a "bastion of democracy and tolerance"; that nazification was an outside political force that invaded the university, destroyed its autonomy, and turned it into a political institution—meaning that the problem was an external one; and that this nazification was carried out by a small number of fanatics, while most professors wanted nothing to do with the regime and even opposed it by maintaining their standards of "objectivity" in research and teaching. The myth prevailed because of the unwillingness of the postwar faculty to face the past and purge the university of Nazi "fellow travelers." Even those who had no love for the Nazis, such as Karl Jaspers, were reluctant to support a thorough housecleaning. Although Remy's argument is exaggerated, he does make the point that the university system was not an exception to the incomplete denazification of German politics and society in the Federal Republic.[9]

Remy emphasizes the failure of the Heidelberg faculty by discussing the most notable exception to this rule: Alfred Weber. In addition to the Nazi flag incident described in chapter 6, Weber refused solicitations for articles from Nazi institutions and leaders.[10] He remained a member of Marianne Weber's Circle, which was not broken up despite the fact that it included those purged from the university. The Circle, in their meetings and writings, were careful to veil their opposition to the regime. Alfred hoped for an allied victory from the beginning of the war, and, in 1943, he made contact with members of several resistance groups.[11]

After the war, Weber was a prominent actor in the "denazification" of the university. He was a member of the Committee of Thirteen, which, with the blessing of American occupation forces, took a leading role in the reconstruction of the university. Of the members of this group, Weber was the most adamant about purging Nazi sympathizers from the university, especially from his own faculty. He was almost alone in believing that there should be no rush to reopen the university if it interfered with the cleansing process. Finally, he did more than any other professor to restore the positions of those faculty members who had been dismissed by the Nazis, a number of them from his own InSoSta.[12]

Alfred Weber's political activities reveal the ironies of his career. In the empire, he created a model that partially departed from those of both the discursive coalition and its cultural opponents. But he could never abandon, as did his brother, the organic structural presuppositions that were so tied to that early habitus. As the polities in which his career developed changed—from empire to republic then dictatorship and to republic again—the monadic organic model remained. This consistency was central to the eclipse of his reputation as a social theorist. At the same time his political activities continually rejected the past, whether that of the state-centeredness and narrow cultural elites of the empire, the middle class antipathy toward the SPD and the proportional representation of the republic, or the entire National Socialist regime. In asking the question, "have we Germans failed since 1945?" he based it on whether they had escaped for the "eternal yesterday."[13] The same could be asked of him individually. And the answer is that as much as he tried and, in his political activity, succeeded, he could never completely escape that earlier habitus. He remains the most complete example of what Kurt Lenk called the "tragic consciousness of German sociology."[14]

Appendix 1: A Note on the Translation

In 1992, I visited Friedrich Tenbruck in Tübingen to get his appraisal of whether a book on Alfred Weber was worthwhile. He responded in the affirmative, but warned me that Alfred's writing style was not nearly as clear as his brother Max's. Having struggled through some of Max Weber's sentences, I almost abandoned the project right there. Hans-Ulrich Wehler was even less charitable:

> The syntactical eccentricity, the unrestrained tendency to connect an accumulation of pallid abstractions with very unbearable expressionistic language from the environs of the George Circle..., the retention of fatally inappropriate expressions...—these raise high hurdles in such a way that the reader must be very dedicated to continually overcoming them.[1]

Needless to say, rendering Alfred's prose into readable English has been a task.

Most theories of translation fall between two extremes: one pole assumes that the text should be rendered in the idiom of the reader even if that distorts the original somewhat; the other pole assumes that the translation should be as literal as possible even if that makes the text less accessible to the reader. I have tried to track a middle course between these two poles. Realizing that one can never capture the original intention of the author, I have tried to make my translations as readable as possible. This was not the easiest endeavor, given the problems of Weber's style. But I have also heeded Walter Benjamin's warning that "consideration of the receiver never proves fruitful"[2] with regard to Weber's terminology. There, I have tried to make it as consistent as possible, especially with regard to key expressions. This sometimes results in rather inelegant prose.

"Fundamentals of Culture-Sociology" has been previously translated, in 1939, by G. H. Weltner and C. F. Hirschman in typescript form and issued by the US State Department of Social Welfare and the Department of Social Science, Columbia University as part of a Works Progress Administration translation project. I have built off that edition, correcting some errors, but more importantly bringing more consistency to the terminology within the text and with other of Weber's texts. I doing so, I believe I more accurately preserve the dualistic character of the German original. Because so few of Weber's writings have been translated into English, I have interspersed longer direct quotations throughout my study. The same guidelines of translation apply to them as to the complete essay.

Appendix 2: Fundamentals of Cultural Sociology: Social Process, Civilizational Process, and Cultural Movement (1920)[1]

Foreword

The topic of the following discussion represents a task that I have pursued and that I first treated in a very incomplete way in lectures at Heidelberg during the winter of 1909–10.[2] At that time, I wanted to contribute to the discussion of the situation of our Occidental culture, one that would move beyond hopes and wishes by employing the accessible sociological insights to provide orientation concerning our present place within the general movement of history. One thing was clear: on the one hand, a closer scrutiny had to reveal the necessity of working out certain basic views of a sociology of culture, and, on the other hand, this had to be applied to history to continually broaden our perspective and to almost infinitely extend the material to be mastered, even where the latter (as happens in every cultural sociology) could only be gathered secondhand. I had hoped, despite professional duties and limited support staff, to offer something preliminary in 1915.

The war and its consequences, which tore me as well as others away from all scientific activity for four and a half years, transformed this and other scientific efforts to a pile of rubble. However, my initial formulation was manifestly essential to, and correct for, our current state of consciousness.

By no means did I intend to synthesize the historical facts into a single picture using the brackets of a general philosophical and historical-theoretical way of thinking with the claim to be offering the foundations of a new view of history in general and, at the same time, a new philosophy.[3] The task I set myself was far more modest. I wished, and still wish, to remain within the framework of sociological analysis, that is, to start with the typical principles of form that govern cultural movement and to typologically analyze the cultural history and cultural destiny of the various great historical bodies in order to apply the acquired results, which could be called the "schema," to the present situation. However, it proved to be necessary to address previous philosophies of history, but only to clarify the concepts of "culture" and "cultural movement." This I did in a lecture on the sociological conception of culture, the basic ideas of which I hope to present here in a better form. This should be followed by a complementary analysis of the principles of form governing Occidental cultural development (its periodicity, forms of expression, cultural dynamic, etc.) and of the sociologically graspable gestalt of its essence, something that was presented in a sketchy form in the lectures of 1909 and 1910. Equally sketchy was the attempt to analyze the cultural-sociological quality of the nineteenth and early twentieth centuries by means of the views that were attained. Although my professional duties prevent me from carrying out this program even in part, I shall, nevertheless, try, in an informal series of essays, to present the previously conceived basic ideas for discussion.[4]

I

It seems appropriate for all sociological approaches to culture to distinguish between different spheres of historical events, namely: social process, civilizational process, and cultural movement.

It is the essence of political as well as economic and social history to examine the destiny of the great historical bodies of humanity, those great unities that interconnect geography, events, and culture, with the aim of clarifying their peculiar destinies by establishing the *concrete* facts that appear essential to the total process. These disciplines regard the Chinese, Indian, Near Eastern-Egyptian, Classical, Arabian, Germano-Romance, and other historical circles, each as a somewhat "corporeal unity" containing a course of events, each as a totality that is connected in that way to action through place and time; and, for the collective destiny of each, they assume the task of collecting the principal data. Accordingly, they seek to embed their description and, in part, their explanation of the major events, the portrait of the great men and the fate of the masses in the form of the economy, in the structural development of political patterns, in the social reconfigurations and other corporeal formations and transformations of historical patterns. Their work is concrete historical morphology.[5] The introduction of so-called spiritual factors and currents does not disturb their essential preoccupation with corporeal destiny. At the same time, the histories of art, literature, music, religion, philosophy, and science—in a word, all parts of *cultural history* that are today separate disciplines (cultural history does not exist presently as a *unified* discipline)[6]—operate in a profoundly different manner and fairly independently of one another.

For them collectively, the *corporeal* formations of the historical patterns of destiny above all are *not* present as an *essential* object of examination or as essential data of development. Their interpretation and explanation of the great cultural emanations and movements with which they are concerned, the spiritual currents and systems of ideas that they seek to uncover in their essence and bring home to us, proceed (insofar as they consider it incorrect to restrict themselves to a mere portrayal of form and content) from the enlightenment of *contexts*—contexts, generally speaking, of "problems" to be solved in the cultural fields on the one hand (problems in the history of philosophy, etc.) and, on the other hand, the principles of technical methodology of the various fields, their development and expressive values (development of painting technique and the plastic arts, laws of harmony in music, laws of language development, of literary styles, and forms of expression, etc.). The result is substantiation of a sequence and rhythm of events that are usually not more closely tested methodologically, substantiation of a conflict of "spiritual currents," styles, forms, and sundry—but always a substantiation of a progression of events which, according to its nature, seems to lie either technically or intrinsically within the principles of the cultural fields *themselves*. These disciplines view cultural history according to the principles of their operation, largely as an *autonomous* sphere of history, whose movement and development they seek to explain intrinsically.[7] The political historian claims the right to weave, somehow, the products of all these individual cultural-historical disciplines into his view of historical events; to place the "spiritual currents and facts" elucidated by the other disciplines into the setting of "corporeal" events that *he*, in turn, elucidates. In this way, he configures his version of the destiny of the great historical bodies into a *general* view, and by combining all these general views, he claims to write universal history.

It is a rather motley and, for reasons in truth pertaining not only to the history of science but also to necessary working technique and methods, *noncontextualized*, loosely and superficially matched mixture of building stones that confronts the sociologist when he finally undertakes to view things *uniformly*. Let him attempt to comprehend any part of historical events, for example, the cultural course, as a *whole*, in the *necessity* by which it grows out of the general course of history and undertake to establish its typical and nomological connection with this general course. Let him, as a cultural sociologist, try to connect *through necessity* the cultural emanations of the Occidental historical world—their essential import, the recurrence or nonrecurrence of their typical forms and aspects—with the fate of the large Occidental community of destiny. Let him try to place these emanations in the body of objective facts (corporeal facts), which the various historical disciplines unearth and which mark history's general course, so that they are bound with the latter in an intelligible and manifest way. In such attempts, he is confronted at the outset, as we have said, by a series of events that are factually discrete and as a whole, in the general presentation of history, only superficially connected. And should he wish to connect these series, the difference between *his* goals and the tasks of the various specialized disciplines will force him to organize his material differently. In accordance with his *goals*, he must conceptually order and imaginatively contextualize the totality of history in other groupings. Thus, whatever facts the political, economic, and social historians have established concerning the external form of history, will necessarily appear to him in a somewhat different perspective. They will come before his eyes as a unified great *social process*, which exhibits *typical* forms and stages of development, despite the great heterogeneity in the different communities of destiny. The major events (wars, revolutions, reformations, and the like) will, in some typical fashion, incorporate themselves in these forms and stages, and the great men will be situated not accidentally but *necessarily* in certain places. *Furthermore*, he will find that this social process is influenced by the *spiritual*[8] sphere, that is, by facts and effluences offered to him by the *cultural* disciplines. When he primarily examines its *core*, he will see it as the form which, under certain natural conditions (geographical, climatic, and other), gives some necessary pattern to the totality of *natural human forces of impulse and will*, which operate in the various communities of destiny and are comprehended as "population." More than *a* gestalt, it is a matter of developmentally changing configurations, which follow, struggle with and separate from one another and, in their struggle, produce the great peripeteiai, the secular historical events. At the same time, he will notice how, in the larger communities of destiny that he observes (and also views as corporeally closed systems), this effluence arises from primitive relationships, residues of aristocratic configurations, which first appear to him on the illuminated historical stage, and passes through *similar* forms everywhere though, to be sure, in totally different groupings. He will observe how the effluence seems to lead via stages of social movement finally to various outlets, to a lasting rigidification of form, to senile decay, or to world expansion of forces, passing through parallelisms to the various ways in which its destiny empties into the universal passage of human history. He will see the Chinese and Indian historical circles—once their natural conditions and direction of development are shown—complete a *necessary* effluence of their social process through the millennia and, finally, yield to that senile torpor in which they remained through the centuries, and in which they still remain today, washed upon by the world expansion of the Occidental historical circle. Likewise, by considering the natural conditions of existence (chiefly the systems of canals and irrigation), he will distinctly recognize the type and direction of social development in the ancient Near Eastern and Egyptian cultural circle, whose early millennia before the birth of Christ he can today reinvestigate by means of unearthed documents. And he will comprehend, as a necessary result of their development, the senile torpor that allowed both to be surprised during the last millennium before the birth of Christ by a new wave of development, that of the ancient Mediterranean historical circle. In the same way, he will observe how the conditions of existence, notably the sea, its commerce and "freedom," led the latter through a given social development—social development always in the widest sense, as pointed out, comprising all that has taken place corporeally within the historical circle—that must bring that circle finally to a type of world-expansion necessarily leading to the senile dissolution of its own forms and corporeality. The historical decline of late antiquity in the time of the Caesars is nothing else but this kind of senile dissolution. And so he will see the Occidental historical circle, which, since the Germanic migrations, pushed the location of the geographical scene beyond that of antiquity to the north, passing through an entirely different, yet equally necessary, development in accordance with the conditions "which ushered it in." This development transmits its corporeality through many evolutionary stages and convulsions, finally to a world-expansion (indeed the greatest there has ever been) actually embracing the entire globe. And now its "inherent" forms also seem to be dissolving, and the circle itself is probably rolling toward something new, either its external decline or the formation of another historical body.

In short, the sociological observer will always see the concrete course of events of the various great historical bodies, their more or less *corporeal* destiny, which is presented to him by the political, economic, and social historians as a social evolution, which is *specific* but, nevertheless, whose content is predetermined by natural conditions, which undergoes regroupings and realignments of *general forms*, runs through a predetermined number of stages, and leads to a predetermined result. In this evolution, the universally given social forces always assume certain coloration, universally given social forms preserve a definite, always specific character and preponderance, universally given processes occur in different groupings and with different results—in all of which a general *principle of social development* works itself out only in various patterns. The major events and upheavals substantiated by the historian thus become landmarks indicating stages of development, or the expression of the peripeteiai bound up with evolutions, and the great men seem to rise as shield-bearers and exponents of the appearance of new periods.

This is the way the sociologist groups the concretely individuated material of the "corporeal" development of the different historical circles, which is supplied by the historian, into a new conceptual form adequate for his mode of thinking—the way he transforms the mass of historic events pertaining to these historical circles into his portrayal of that sphere, which I propose to call the *social process* of humanity.

II

As stated, the sociologist will recognize that this social process, which, for him, is *primarily* born by the natural forces of human drives and will, and which is brought into a certain form and direction by certain natural conditions peculiar to each historical body, is *also* determined by factors that are established by the other group of historians, the "intellectual historians": "ideas," "spiritual currents," artistic perceptions, religious convictions, and so on. He must at first be indifferent to their closer *dynamic* relation to the stages, peripeteiai, social configurations of form, and all else pertaining to "corporeal" development, their causal influence on this development, the prius and post of the form and content of the "spiritual" and corporeal spheres.[9] In any case, he sees a spiritual-cultural sphere existing as a *totality* in each historical body, alongside the "corporeal." And, no matter what he may think of their mutual interaction, he also discerns in this spiritual-cultural sphere, which is perceived as a *whole*, regularities that stand in a yet unexplained connection to the corporeal social process. He discerns in it a blossoming and an aging, a destiny that parallels that of the "cultures" which are present in the various historical bodies, a somehow predetermined appearance of successive developmental stages; a characteristically recurring rhythm of productivity; an emergence, variegated yet exhibiting certain regularities, of the different *aspects* of cultural expression (religion, philosophy, art, and within art: music, epic, lyric, drama, painting, etc.) and *types* of expression (classic, romantic, etc.); a characteristic recurrence of great religious movements and related currents of ideas under similar conditions in the social process of the various "bodies." In short, he discerns a spiritual-cultural development in the various historical bodies that is related in some fashion, or at least is somehow parallel to their social process. He sees himself compelled to view this spiritual-cultural development also as a *unity*, a second sphere of historic events. For this purpose, he has to lift out the disconnected facts that are presented to him as parts by the different specialized disciplines and order them by placing them in a *totality* of historical movement, as a total process occurring in the various historical bodies side by side with their social process. He is thereby tempted—in fact, it is his essential duty—to clarify the dynamic relationships of these spheres (which are sociologically perceived as unified) to one another within the various historical bodies.

But when the sociologist seeks to go about this task of scrutinizing the spiritually cultural sphere, he encounters something peculiar. He notes, namely, that something imposes itself between the social process and the intrinsically *cultural* parts of this spiritual-cultural sphere with their various aspects and forms of expression in religion, art, and so on—a spiritual intermediary realm that stands in a much closer and clearer connection to the form and course of the social process than do the intrinsically a potiori cultural phenomena (the emergence of religions, systems of thought, periods of art, etc.). This is, namely, an *intellectual cosmos*, which offers the social process the technical means for its forms and configurations and, likewise, appears to be one of the foundations of cultural-phenomenology. To express it more accurately: his experience is that the spiritual-cultural process of the various historical bodies, which he has lifted out of the isolation of the various specialized disciplines and viewed tentatively as a unity, is really in its essence, in its developmental phenomena, its development, as in its relations to the social process, no unity at all but a *duality*, and that it carries within itself two entirely different spheres of human historical development, which cannot be properly unified according to a single perception.

What is revealed upon closer scrutiny is that in every great historical body this "spiritual-cultural process" contains a threefold entity. First, the most intrinsic and "purely spiritual" element of this totality is the *development of* a popular *consciousness*, which proves itself to be the kernel of the purely spiritual process of growth and aging of the historical and cultural bodies once these are viewed from the spiritual-cultural angle. In each of the great historical circles that he can sufficiently investigate, those of China, India, Antiquity, and the Occident, the sociologist observes that the development of consciousness proceeds typologically, from primitive stages where the forms, through which the world and one's own ego are seen, approximate those of the today's aboriginal and half-cultured peoples, to an ever-increasing critical reflecting about existence. This development discards the totemistic and then the mythical representations, or, at any rate, gives them a place in existence that is no longer naive but rather determined by reflection. He will watch it advance further from a purely empirical position vis-à-vis the world and the ego to one that is in some way scientific or, at the least, intellectually formed, the advance determined by some kind of intellectual abstractions. He will see how these abstractions are further developed, how, at a certain stage, every historical body harbors some rationalized system of observing the world, which can further elaborate and change itself, into which not only external experience, "the world," but also one's own "ego" (one's emotions, drives, and immediate spiritual images) is "worked out" and given definite though varied forms of an intellectual, systematized way of perception.

The sociologist discovers that in all the historical bodies in which he observes this process, it is intimately connected to a second and a third process within the same "totality." The second is an increasing spiritual domination over nature that (parallel to the intellectualization of the image of world and ego, as another side of the same development) presents an intellectual formation of the *practical*-utilitarian cosmos of knowledge, of experiences, and cognizance of life, and which also yields to the tendency toward intellectually systematized form. Moreover, it remains a self-contained process, even through the various stages of the various historical bodies. Finally, he sees as the third spiritual process something that is none other than the actualization and concretion of this second intellectual cosmos to which it is connected—the transformation of this system of practical knowledge that is constructed there into something completely material through the cultivation of an apparatus of tools and methods, principles of organization, and so on, which shape existence into concrete formations.

At this point, the whole spiritual sphere, borne internally by the development of rationalizing consciousness and projected outwardly in both the above-named senses, extends directly into the social process, codetermining it through this configuration of the technical apparatus. In short, the sociologist beholds, as something particular and self-contained, a great *process of the rationalization* of existence, which possesses only different aspects of expression and pervades all the great historical bodies. The process codetermines the latter's forms and its emanations affect the *internal* existence as well as the *observational* and *practical* treatment of the external existence. This process of rationalization has its own laws of development, its own necessities of growth and its own conditions of stagnation. Manifestly, it is a completely different entity from the emergence of religions, systems of thought, works of art, and cultures. It is a unique and vast sphere of development, which stands in a quite different relation to the social process than these do. Once seen as a unity, this sphere divides the previously intuited unity of the spiritual-cultural sphere into a "duality."

This process of intellectualization and rationalization that pervades the historical bodies, and the intellectual cosmos that it displays everywhere, its unity that is reflected in its three expressions (inner intellectual elucidation, intellectual formation of knowledge, and the intellectualized external apparatus), its operations, forms, and patterns—one has to view all these as an *especially* large sphere of historical happening that is to be distinguished *as much* from the sphere of the social process *as* from cultural movement proper and is to be intuited and investigated in its consequences and particular regularities as a unity. On the whole, this unity was not emphasized in this way by previous historical and sociological accounts.[10] I propose to call it the *civilizational process* and to demarcate it and its sphere sharply and fundamentally from both the social process and the sphere of cultural movement. The latter is *also* embedded in the social process of the great historical bodies but, as we shall see, is related to it quite differently from the civilizational process, is governed by entirely different laws of development, is of an entirely different essence and has an entirely different place in the course of history. I propose, for the purposes of the cultural-sociological approach—perhaps for the sociological approach in general—to resolve the process of history so that the "corporeal" element in its development (that which we have named social process, *primarily* the realm of natural forces of drives and will and their formations) can be posited separately and considered, first, as being influenced by the civilizational process, humanity's sphere of rationalization. Then one can ask from the other perspective: how the cultural movement proper stands in a context with both and with their interaction; whether it grows in some recognizable fashion out of their interaction and in their forms and patterns; whether and to what extent it proceeds independently of them, and how much it reacts upon both. I am proposing this kind of trichotomy because this is the way to attain a unified sociological representation of the course of history and especially (as I believe and intend to prove) a sociological analysis of its cultural phenomenology.

III

Civilizational process and cultural movement are, as I said, intrinsically diverse; they have completely different forms and laws of development; and they appear before us in the general course of history as mutually exclusive phenomenologies.

The civilizational process—in its various composite parts: its picture of world and ego (macrocosmos and microcosmos) formed by the intellect, its pragmatic cosmos of intellectual knowledge, and its intellectually formed equipment for mastering existence—may reach entirely different levels in the different historical bodies. It may clothe its picture of the world in considerably diverse forms of expression, but in every historical body it always builds, stone by stone, a cosmos of *knowledge* whose three indicated parts are merely its different fronts of expression and whose evolutionary formation, once set in motion in a certain direction, proceeds nomologically, just as the construction of a building is governed by laws of inherent causality. Whatever is produced, the whole and its parts, is not "created" but "discovered," that is, already *there* (given the direction of the intellectual movement) before they are found. They are *preexistent*, and in regard to their course of development, are, so to speak, only drawn into conscious human existence, into the elucidated sphere of being with which man surrounds himself. This applies to the entire cosmos of practical knowledge in the natural sciences, as well as to every individual "discovery" of natural science. It applies as well to the general theoretical system of knowledge and to every individual epistemological insight. But it also applies to the entire technical apparatus of existence, to every individual tool, machine, and methodical principle of work and organization that is found. The propositions of Euclidean geometry are "present" prior to "discovery"—otherwise they could not be discovered at all; and the same is true of the Copernican formulas for planetary motion and Kant's a priori to the extent that all these are "correctly" discovered and formulated. The same is so for the steam engine, telephone, telegraph, ax, shovel, paper money, division of labor, and the whole body of technical means, methods, and principles concerned with the mastery of nature and human existence. All "objects" of the practical-intellectual cosmos of our existence, all that we already possess or that are still to be acquired are in essence there, "preexistent," before we are able to draw them into the conscious sphere of our existence and utilize them. The total civilizational process that actualizes this whole cosmos and supplies us with all its "objects," including the discoveries of the purely spiritual world, merely discloses an *already present* world, a world that for us as humans is *universally* present, and renders it progressively accessible. This world is universally present for all people, and every part of it *is "valid" for everyone.* This is proved—I shall soon touch upon apparent deviations—by the fact that if the objects of this world (its spiritual and corporeal concretions) are discovered in some historical body, no matter where, and are drawn into conscious existence, then by their entrance into human existence they spread throughout the world as though their movement were self-evident, necessary, and wavelike. And they are employed overall in other historical bodies, to the extent that the social process is sufficiently developed in these bodies to receive them and to the extent that the consciousness of the soul has sufficient elucidation for them "to be seen"—provided, of course, that given the communication network they *can* be seen at all. The universality of technical discoveries is well known. But this universality is not restricted to the "technical" cosmos of civilization whose material and spiritual objects, methods, and means—from the knowledge of the working of metals and the use of fire to today's means of communication and production—have always spread over the world more or less in a flash as though through electrical elucidation, both in periods of universal communication or isolation. This also applies to the cosmos of intellectual knowledge. The latter's insights in mathematics, astronomy, the natural sciences, and so on, may spread more slowly at times, since their reception depends on the level of consciousness attained in the different historical bodies and since many of their practical applications (measurement of time, accounting, etc.) are not feasible due to the social organization. But this does not prevent them from finally prevailing everywhere in the same measure. And the same *universality*, with certain modifications in the form and manner of expression (addressed below in more detail), also holds good for the exposition of new parts of the intellectually shaped *picture of world and ego*, the intellectual results of the elucidation of consciousness—the clarification, so to speak, of the inner front of the preexisting civilizational cosmos. When viewed as a unified historical picture, the phenomenology of actualization and development of the civilizational cosmos (in both its practical and theoretical aspects) means that the great historical bodies, which are widely divergent from one another in their *social* and *cultural* development, build entirely upon one another in their *civilizational* development and work, on the working out of something thoroughly unified as if according to a provided plan. Indeed, so viewed, the entire course of history is really the process of the unified exposition of the universal civilizational cosmos of mankind, which characteristically comes to fruition only in the breaks, stages, and discontinuities imbedded in the destiny of the different historical bodies. The historical circles of the ancient Near East-Egypt, Antiquity, the Arabian lands, the modern Occident, and China and India (which are more loosely connected to the former), no matter how acutely they diverge from one another in their *historical* course, *social* development and cultural movement, are all, from this perspective, only links, to a certain extent auxiliary factors, which permeate the entirety of history and (in a series of stages) bring about the continuous, logical exposition of the civilizational cosmos that is common to all mankind today.

The *technical* parts of this civilizational cosmos, as they appear in their present rational form, are first visible historically in the organization of tools and labor by the Egyptians and Babylonians as far back as 3000–4000 BC. Having evolved in correlation with the historical circles of India and China (the details of which are not known), this technique became the foundation, not only of the whole civilized technical apparatus of the Classical and Arabian historical bodies, but also, through them, of the present Occidental body. The latter, on the basis of what was already created by the entire world, has taken the lead in technical invention since the fourteenth century and, since the eighteenth century, has produced the present technical apparatus of world civilization.

In like fashion, the *spiritual* parts of this world-civilizational cosmos, mathematical, astronomical, and natural scientific knowledge, apparently had their intellectual elucidation in the enormous depths of the first two historical bodies on the Euphrates and the Nile. They were then brought into sharper relief by the Classical, the Arabian and, from another side, the Chinese historical bodies. Their further development has been taken over by the Occident since the sixteenth century and carried through the famous "Era of Discovery" to the present universally prevalent conception of the world based on mathematics and the natural sciences, a conception which is "valid" for, and universally accepted by, all mankind.

The intellectual *"cosmos of consciousness"* that, despite its continued expression in various forms, is, through its content, presently the civilizational common property of humanity—the "picture of ego and world" as belonging to an intellectually perceived sphere of unity—first seems to have attained conscious elucidation in the Brahmanic wisdom of the Indian historical circle. It then became an object in the Classical and Arabian as well as the Chinese historical circles, and finally, in the Occidental philosophy of the eighteenth century (Kant!), it received those intellectual principles of form that seem to show the limitations of its capability of elucidation while contemporaneously collecting the different forms of elucidation of the various historical circles and generalizing them, insofar as they possess an *intellectual* content.

In this gradual emergence of the preexistent spiritual and material civilizational cosmos of humanity from the darkness into the light of human collective consciousness—sketched here only in an amateurish and inadequate manner—it is of little import, a "minor accident" so to speak, if certain acquired knowledge and clarifications get temporarily lost through historic contingencies, above all through the way history has of telescoping the formation of historical bodies that become the carriers of the process of elucidation. Take, for example, the knowledge of the Copernican picture of the world, which, after its discovery in Greco-Roman antiquity, slumbered in the lap of history until its independent rediscovery by the Occident since the sixteenth century. Equally irrelevant to the nature of the whole process is the possibility that in the exposition of the "technical cosmos," certain technical means of civilization "accidentally" discovered somewhere initially remain unused until their rediscovery somewhere else, when they suddenly receive enormous significance and a universal, practical application. Thus, although the early discovery of the mechanical clock or the engine in China was not followed by social application, their "discovery" again in the Occident ushered in the great technical revolution of modern times. These are not changes in the essence of development but the "jests" and curling arabesques of the process that result from its being embedded in the social and cultural movement.

And finally, it is irrelevant to the essence of the civilizational process as a gradual emergence of a spiritual type of unity that the development of *consciousness* underlying it is widely set back in the early "history" of the various historical bodies, always having to begin anew from a relatively primitive state in a part of the world. This was the case with the development of the Classical consciousness, which succeeded the Near Eastern-Egyptian. (The migrating and invading Greeks were obviously barbaric compared to the Cretan-Mycenaean historical circle that they encountered as an offshoot of the Near Eastern-Egyptian.) The same is true for the development of the Arabian consciousness, which succeeded the Classical, and for the Occidental historical circle, which succeeded both of them. This means simply that where there is an influx of new peoples into the general civilizational cosmos of mankind, the "subjective" civilization or "civilized quality" of the new populations must always ascend anew the stages that have already been discovered and traversed by others within the general objective and subjective civilizational cosmos. Here, by the way, the climbing and reaching for old subjective heights of civilization is always facilitated considerably by the fact that the most essential *objective* elements of civilization are taken over by each new historical body and, with them, those that are of supreme importance for the acceleration of the subjective process of civilization, the subjective intellectual elucidation of consciousness and the mastery of existence through consciousness. When, for example, the Classical historical body took over from the Near Eastern-Egyptian not only the technical apparatus and the principles and forms of division of labor, but also coined money, mathematics, and astronomy, it thereby took over the ultimate elements of "objective" civilization. These made possible, immediately, a measurable intellectual domination of existence. They facilitated enormously the rationalistically conscious mastery of the "inner" and the "outer" things of existence. These contributions were no little thing. They gave rise to the enormously rapid elucidation of consciousness and civilizational development that were brought to consummation with the "Greek barbarians" a few centuries after their incursion through the Doric migrations; just as on the other side they also influenced, *in terms of content,* the remarkable early rational formulation of their view of world and ego. But this is only in passing. The same thing can be said, for example, of the effect of the transmission of the Classical money accounting on the Occidental historical circle after the Germanic migrations with regard to the development of consciousness and civilization of this historical body, which had, in large measure, again become "unconscious" and block-like and expressed itself only in primitive social forms. We find the general "money accounting" and, therefore, the beginnings of "a calculating spirit" in the Germano-Romance historical body—as is evident from the *leges barbarorum*—long before the development of a money-exchange economy of any essential significance.

There is no doubt that, in *this situation*, "subjective civilization" is set back for centuries through an influx of new peoples in large numbers into the general civilizational cosmos, through the formation of a new historical body, through the dislocation of the historical process around a new center of gravity, that is, into a new geographical setting in which the historical body must then grow and finish its social and cultural development. And *subjectively* viewed, a stage of antiquity must always recur, followed by a medieval and a modern period. Consequently, the *subjective* civilizational process of all humanity undoubtedly presents a picture of constantly recurring darkness in certain "areas" in which humanity is historically integrated, until gradually the earlier elucidation reappears and is then surpassed. Undoubtedly, however, the preservation of objective civilizational elements and subjective enlightenment in the *other* undisturbed historical "areas" creates the means whereby the setbacks of single parts can be speedily recovered and then the general enlightenment again pushes forward from this or that area. And undoubtedly this general enlightenment is the logically causal (even if it proceeds in an interrupted and splintered series of stages) exposition of a great *unity* valid for all of humanity, its universal, objectively and subjectively preexistent civilizational cosmos.

Whether an aspect of the process of elucidation will actually be executed depends on the specific internal attitude (I shall not as yet use a more specific or fundamental term) of the various great historical bodies and, perhaps (as is recently contended), also on the psychic orientation of their populations, which will be discussed shortly. The ancient Near Eastern-Egyptian body, in accordance with *its* attitude, developed the foundations of the practical technical aspect, but on the "theoretical" side cultivated only the purely quantitative parts that were necessary for the immediate mastery of existence (astronomy, measurement of time, accounting, etc.). In contrast to this, the Classical body, in accordance with *its* specific attitude, did not "see" at all the technical parts of the civilizational cosmos and simply ignored them without special interest. (Except for the arch, no technical invention of Antiquity is worthy of mention.) In accordance with its field of vision, it turned exclusively to the intellectual and theoretical front and, so, laid the foundations for mathematics, the natural sciences, philosophy, and all the other disciplines that we now call "sciences." Whereas the Indian historical body, with its remarkable adaptation to concrete existence, virtually ignored *everything* else and chose as its almost singular cognitive task (which it accomplished with great success) the religiously embedded philosophical elucidation and penetration of the innermost cognitive domains of the image of world and ego. It is very much the normal procedure for every historical body to clothe cognitive acquisitions (those introduced here as well as all others, especially the innermost philosophical ones) in forms in accordance with their "attitude" and means of expression. These forms do not always immediately reveal the universality of these cognitive elements and impede their general expansion and application, especially when these elements occur mixed with extracivilizational elements and enclosed in religious and metaphysical systems of thought, as can be seen especially in the "epistemological" inferences of the Brahmins. A further result of this is that the consciously or unconsciously applied apparatus of representation and conceptualization (which always contains a definite system of mathematics, i.e., a definite structure of temporal and spatial representations) sets completely different limits in different historical bodies according to its quality of elucidation in their content. Without the "idea of function," which appeared first in the Occidental circle, not only all higher mathematics, but the whole of modern Occidental knowledge, could not have been built up. The same relation exists between the Euclidean idea of three-dimensional space and the whole knowledge of Antiquity, and between the Indian idea that corporeal being is only "appearance" and all Indian philosophy. But it was a distinct misapprehension to claim or, at least, to suggest that the "cognitive elements" (in our terminology, the exposed parts of the intellectual civilizational cosmos) are therefore only "symbols of the soul" of the various historical bodies, valid only for them, and that there existed, for example, an Occidental-Faustian, Arabian-Magian, or Classical-Appollonian mathematics whose truth and application were correspondingly limited to those bodies. The exposition of Euclidean geometry may have been a result of the "Apollonian soul" of Hellenism—that will not be disputed here—and it may also have been first presented to the world in Hellenic form. But its content of truth and knowledge is, in the human sense, *eternal*, that is, universally valid and necessary for all humanity. The same is true of the cognitive content of the "Faustian" infinitesimal calculus and all its consequences, or of the Kantian a priori or of the Indian opposition of "essence and appearance." Accordingly, that which Kant, in his test of the formal premises of knowledge, excluded from the sphere of pure empirical knowledge and labeled metaphysics, must once and for all be excluded from the temples of universal *"knowledge,"* from the temple of *civilizational* knowledge, and therewith from the clarification of the universal preexistent cosmos of civilization, its theory, and its practice—but not from the temple of "truth" in general! For we shall meet these metaphysically or religiously conditioned parts of the "spiritual world of knowledge" of the various historical bodies again elsewhere, in their worlds of *culture* and cultural movement. And it will be evident that they possess in this world—no matter how slight their civilizational, that is, universally human, valid, and necessary content of *knowledge*—a tremendous *cultural* and, yes, *psychic truth*-content which constitutes the content and essence of the cultural emanations. But more of this later.

Let us now summarize: The phenomenology and form of appearance of the civilizational process is that of the logically causal mode of development in stages, the accumulative elucidation (although in various splintered pieces) of something preexistent and latent in all humanity, and the disclosure of this as something universally valid and necessary. In addition, the civilizational cosmos is an intellectually formed cosmos of universally valid and necessary things that cohere internally in all their parts and that, viewed from the practical side, are equally and universally useful (practically correct) for human ends. Viewed from the theoretical side, they are equally inevitable (theoretically correct), and are equally immediately evident (a priori correct) in the elucidation of the picture of world and ego. This cosmos is the epitome of humanity's increasing enlightenment. Its exposition, therefore, proceeds by the laws of *logical causality*. At every step in this evolution the terms *correct* or *incorrect* are applicable. And its exposed and elucidated objects have the character of universal validity and necessity and spread with the speed of *lightning* through the cosmos of transportation, for the very reason that they are preexistent for all humanity.

IV

Exactly the opposite characterizes *cultural movement* and everything that originates or is located within its sphere. This sphere produces no cosmos of universally valid and necessary things. Rather, everything that arises here remains, by its very *essence*, confined within and internally connected to the historical body in which it arises. No objective cosmos arises, but rather a psychically determined aggregation of symbols. The Chinese culture is one such independent world of symbols, with its own runic writing and its own ultimately nontransferable content, as are the Indian, the Egyptian, the Babylonian, the Classical, the Arabian, and the Occidental cultures. They are all different cultural worlds with differences in all that is truly *cultural*. It is impossible to separate Hellenic culture from the Hellenic historical body, to appropriate, transplant, or duplicate its content—despite the often-repeated attempts to do so with its essential parts, its fine arts, its drama, and its philosophical world of ideas. Every renaissance—and there have been many *attempted* renaissances of Hellenic culture, from the Augustan in Rome and the Greco-Buddhistic in the Gandhara region to the Italian, the Empire renaissances and others—leads to something completely different from a revival of cultural Hellenism, even though certain external forms were appropriated each time and a similar *content* of psychic *redemption* was repeatedly sought. Every time the content of psychic redemption as well as forms of psychic redemption that appear concretized in works of art and ideas, that is, the newly created cultural world, is thoroughly different from the Hellenic; and the alleged renaissance is, in reality, a *new creation* of something else. The same holds true for the appropriation and dissemination of the contents of redemption that are *purely religiously* formed. In the spread of "world religions," one apparently (but only apparently) is presented with something similar to that which takes place with the spread of the content of civilizational knowledge, namely, their release from confinement within the historical bodies in which they developed and a universalization for all, or at least large parts of, humanity. The spiritual and psychic universalization of the world religions—Christianity, Mohammedanism, Buddhism—even within the boundaries where they were successful, is an illusion. Under closer examination, it may be nothing more than the result of the military expansion of the religion's original historical body—for example, the spread of Mohammedanism is almost concomitant with the final expansion of the Arabian historical body, which had become Mongoloid. A second possibility is that this universalization is a "transvaluational" renaissance in a different historical body, that is, essentially one of those "new creations" we have already seen in art—for example, the spread of Buddhism to eastern Asia. In the case of Buddhism, there is, in large part, not even a similarly directed yearning for psychic redemption. Hence, the "Mahayana," which supplied the raw material for, and was further developed in, East Asian Buddhism, is really an entirely different religion, which was not cosmological but rather filled with a content of subjective beatitude, something essentially alien to the true Indian Buddhism still extant in Ceylon. It applies the latter's intuitive forms, but has, in all its various guises, a different psychic content.

Or finally, in this apparent universalization, the powerful expansion of the historical body and the newly creative transfer of values may combine, as in the case of Christianity and its world expansion. Born as a phenomenon of the psychic old age of Classical antiquity, Christianity was reborn as something completely different in the Germano-Romance historical circle at the time of its *inner* acceptance by the new young world, which did not begin until the year 1000. Since then, it has been not only different in its dogma, but also in its very *essence,* from Oriental Christianity, whose spread into Russia also led to a whole series of actual new creations. And here, as there, it experienced, again and again, renaissances (called "reformations"), which, in the different historical bodies, always led to new creeds (Troeltsch quite correctly expresses the opinion that we should call them new religions), to the formation of new sects, and so on, with completely different content and apparently quite different forms of expression. Christianity in its various forms spread over the "world" within the realm of the expanding Occidental historical body over the same area, and has only moved beyond this since the eighteenth century. But even this alleged "universal religion" of humanity, and notably this, is today a conglomeration of many different religions that coexist with and succeed one another. Each has *equal* psychic content of truth for its respective historical body, in which it is the expression of the current psychic situation and to which, in actuality, it is confined totally in its essence, content, and expansion.

Moreover, the religious, psychic expression of culture, almost everywhere, externally arrays itself in "categories of knowledge." It presents itself as "revelation," as "comprehension," as "certain (immediately intuited) conviction of which one can see nothing" and seeks, through this equation of "life experience" with "comprehension of the invisible," to usurp universal general validity and necessity, to evangelize, to convert, and, especially in the case of Christianity, to persecute and burn all those of different faith. But all this merely covers the underlying fact of the struggle of expressions of the souls among one another, which differ in essence because they are bound to psychic adjustments to the world of the different historical bodies in which they are enclosed.

What is true of religion is ultimately true of the content of the metaphysical ideas of all philosophical systems, which is always purely and simply a cultural expression of a particular historical body. It is utterly impossible to convey to the Occidental, or any other body, the intrinsic content of Indian metaphysics, its belief in the transmigration of souls and its longing for dissolution of individual essence. If we attempt this, we arrive at Schopenhauerianism or theosophy that, although they may externally apply the same or similar forms of concepts and representations, completely transform the content that they embrace and express. Likewise, it will never be possible to give Hellenic Platonism any universality. It has undergone numerous renaissances in the form of Neoplatonism, Renaissance Platonism, German Idealism, and so on, each of which represents a completely new creation in essence and content.

All cultural emanations, religions, systems of ideas, and art creations are in complete opposition to all civilizational elucidations; they are confined, so far as their truth-content is concerned, to the times and historical bodies in which they arose. Their transfer to other times and other historical bodies is always a mere transfer of their expression and psychic *values of redemption*, a *transfer of value* leading to the so-called "expansions." This has absolutely no connection, however, with the logically causal expansion of the elucidated parts of the universal civilizational cosmos.

Thus, all cultural emanations are always *"creation."* They bear the distinguishing sign of all creations, the character of "exclusiveness" and "uniqueness" in contrast to all things exposited by the civilizational process, which always possess the character of *"discoveries"* and thereby of universal validity and necessity, the exposition of something that was already existent.

Correspondingly, the phenomenology of cultural movement, the type of development in the sphere of culture, differs radically from that of civilization. In the latter, as we have seen, there is a "development," broken, of course, and subject to historical contingency but nevertheless proceeding in stages, a *unified process of elucidation* passing through the whole history of humanity and leading to a definite *goal*: the total elucidation of the preexistent. In the cultural sphere, on the other hand, we have *protuberance-like* eruption of productivity here and there in an apparently inexplicable manner, which suddenly brings forth something great, totally new, unique, and exclusive—an incomparable "creation" standing in its essence in no necessary context with other things. And if we observe and attempt to establish certain regularities and contexts, we do not have "stages of development" but disconnected periods of productivity and nonproductivity, periods of decay and stagnation, sudden reversals, antithetical "temporal currents" that struggle with one another. We have, before us, not stages, but forms of expression of new *psychic situations*, a rough sea, sometimes cresting sometimes calm, moved by this or that "psychic" wind, but never "flowing to" somewhere, never striving toward a "goal." Insofar as we can substantiate "developments," they concern only the *technical* means for the *expression* and elaboration of culture, the somehow contextual sequence of naturalistic, classical, romantic, and baroque types of expression in the various disparate periods of productivity, the alternation of more emotional and more rationalized forms of expression of cultural accomplishments (religions, works of art, etc.), the dissolution of mythically veiled expressions by unmythical ones with the aging of the various historical bodies, and so forth. In short, one can observe this "development" essentially not in the content but in the surface movements, which, however, operate within the culture of every historical body independently and separately from the other, as if in another world.

In the cultural movement of the various historical bodies, we really have before us the configuring from within of totally different "worlds," which arise and disappear with these historical bodies, which are unique and exclusive throughout, and which are essentially different from the extension of the uniform cosmos brought forth by the civilizational process.

Whereas one can apply "intellectual" concepts and representations, modern scientific concepts, to the objects of the universally valid and necessary civilizational process and cosmos and thereby configure for our consciousness a picture of this process and its consequences, the objects of cultural movement and the various exclusive and unique cultural worlds can only be approached through the formation of "historical" concepts, concepts and representations of "unique essences." And consequently, for the sociological examination of the cultural worlds and cultural movement, it can only be a matter of elaborating *typologies*, that is, the comparative exposition of a recurring *phenomenology* of the surface appearance and an attempt to bring this phenomenology, and the *unique elements* within its compass, into some kind of elucidating context with the general civilizational process of humanity and with the social process of the various historical bodies. This approach, which must be further developed, is the task of cultural sociology.

V

We can now approach the inner essence of "culture" as opposed to "civilization," their dynamic interrelation and the relation of both to the social process—the real question of the principles of cultural sociology.

No matter how one explains the origin of the civilizational element in the historical process of humanity, the slow and gradual exposition of the delineated cosmos, this cosmos is certainly *humanity's* most essential *resource in the struggle for existence*, a spiritual as well as material resource. Humanity achieves the domination of nature through the intellectualization of the stuff of its experience. In fact, through the resulting exposition of that preexistent civilizational cosmos, humanity establishes, between nature and itself, an intermediate realm composed of the contents of consciousness, knowledge, means, and methods. Through this realm, it tries to master nature, to efficiently organize its own existence, and to revamp and expand its natural possibilities. Indeed, this civilizational cosmos *is* nothing more than this *expedient and useful intermediate realm*, and the range of its universal validities and necessities extends just as far as these expediencies and utilities have any meaning. The civilizational cosmos appears as the concretization of a world of concepts and representations, as the evolution of a view of nature, world, and ego, the purpose of which seems to be the spiritual and then the practical mastery of nature, world, and ego by subjecting them to the light of the intellect.[11] It is the *"picture"* of nature, world, and ego adapted to the struggle for existence, even if not produced by it. And the innermost structure and formal elements of this picture, its aprioristic forms of intuition, its categories, and its synthetic mathematical "judgments" appear a priori as the instrument that somehow has arisen in the human spirit, in order to slowly construct this cosmos of mastery. Since the *fundamentals* of the struggle for existence are alike for all people wherever they might live, it seems evident that the inner spiritual instrument in the struggle—these categories of perception—is also alike for all, and that the "picture" of the civilizational cosmos exposited by this instrument must be common to all, that is, it must possess universal validity and necessity. That said, its remarkable apparent preexistence ceases to be miraculous; it is the consequence of the basic categories of perception developed everywhere in the same way. The civilizational cosmos is nothing but a "world picture" slowly constructed and elucidated by and through these categories. It is the "aspect of nature" that they "*fabricate.*" This view of nature is eminently suited to the purpose of mastering nature and existence in general and creating the external "realm of domination," that is, the civilizational apparatus of existence, because it grew out of categories, which, in turn, appear to have arisen for just this purpose.

This is the true essence of the civilizational cosmos, which therefore represents the great realm of *existence that is expediently and usefully elucidated and formed*. This cosmos, however, does not move even one step beyond the point of view of expediency and utility in its *formation* of existence.

In contrast, the *cultural* formation of existence has nothing to do with expediency and utility. That which influences existence in religions and systems of ideas and is mirrored in works of art and "forms" springs from a province of quite different categories and intuitions, from that of the *soul*. The *psychic* elaboration and *formation* of the stuff of existence is antithetical to the civilizational, that is, intellectual, elaboration of it. The nineteenth century is largely to blame for demolishing, to a certain degree, the province of the soul, the psychic sphere of humanity, as the ultimate and deepest sphere of *essence* for the knowledge and perception of the historical process. This most intrinsic sphere of essence—in light of which everything else in existence is seen, or ought to be seen, as nothing but expression, form, incarnation, phenomenon, metaphor, or symbol of a "psychic essence"—has been made invisible to the historical, philosophical, and, consequently, the prevailing sociological perspective through the concept of "spirit"—above all, the Hegelian "objective spirit." This concept of objective spirit connected the *intellectual* elements for the *mastery* of existence with the *elements of psychic expression*, thus actually rolling intellect and soul into one,[12] and hopelessly confusing civilization and culture. Culture, however, is simply the soul's aspiration and expression and, thus, the aspiration and expression of an "essence" lying behind all intellectual mastery of existence, of a "soul," which, in its aspiration and striving for expression, does not concern itself with expediency and utility but only with the penetration and formation of the stuff of life. This soul exhibits a sort of likeness of itself, and through this likeness, through the imposition of a form upon the stuff of life, or beyond it, facilitates its own "redemption." All culture is nothing other than this striving for redemption by the soul of the various historical bodies. It is the soul's attempt to achieve the expression, pattern, image, and form of its own essence—either to shape the given stuff of existence or, if this is impossible, to escape from it and to seek a transcendental existence as a realm of formation and redemption.

This means that if the social process is the *"corporeal"* in the development of the different historical bodies, then the civilizational process supplies it with the *technical* means to construct this or that expedient or useful form of existence. But for the cultural movement, all this is merely the *substance*, the stuff, that it must psychically elaborate and transform into an expression of the living soul in the various historical bodies as well as give form to each configuration of essence. Thence follows the concept of culture as *the form of psychic expression and redemption within the given material and spiritual substance of existence at any one time*. And already this yields something preliminary concerning the rhythm of cultural productivity in the various historical bodies and the reciprocal dynamic interrelationship of social process, civilizational process, and cultural movement.

VI

The social process of the various historical bodies, that is, the corporeal historical process viewed sociologically along with the bonds and struggles of the natural powers of humanity that are formed within it, passes through various stages. It passes from simple to complex forms of *life aggregations*. In this development, it undergoes complete social regroupings, broadening and narrowing of horizons, rigidifications and dissolutions of its social forms. All of this is taken as lived experience and as the stuff of formation from the standpoint of the soul that dwells within it. It undergoes partial restratifications of the stuff of existence and complete *reaggregations of the elements of life*, which present existence as a new totality of lived experience. On the other hand, it also passes through longer or shorter periods of lasting rigidification, times of "rigidity of the *aggregation*," of pure repetition of lived experience, which for generations or even millennia perhaps, offer the soul nothing but the same lived experience and material for formation. In addition, the civilizational process, through the presentation of new technical methods for the formation of existence, new knowledge and horizons, is one of the most essential factors in fostering the further development of the corporeal configuration of society from more simple to more complex forms. It allows the continued cultivation and the dislocation of the corporeal aggregation of life, and is able to bring about, by inventions and discoveries, great and sudden changes in the aggregation, great reconstructions and completely new aggregations. Furthermore, by stagnating in some historical body, it can contribute to its corporeal rigidity and further its dissolution through aging. It creates, therefore, *new external aggregations of life in conjunction with the natural forces of the social process.*

The civilizational process can also create, however, a purely *spiritual* reaggregation of the elements of life *without* such a transformation of the *external* aggregation of life by bringing forth only new, purely *spiritual* elucidations or by penetrating from without into some historical body. And it does *this* by setting itself spiritually in a new connection with all the elements of existence, completely independently of the corporeal transformation of existence. This aggregation can be perhaps equally as great as any brought about by a corporeal regrouping of society. If hitherto the "world" was seen as a saucer shaped disk, over which the "canopy of heaven" arched like a foreshortened bell, and now the Copernican view of the universe with its perspectives of infinity suddenly appears everywhere instead and is universally *comprehended*, then a new aggregation, a spiritual revolution and reordering of all elements of life begins, which is just as great as, or for the "soul" perhaps even greater than, any other corporeal transformation of the elements of life. The same sort of thing must occur when one suddenly no longer sees the external world as something independent of our ego and its forms and conditions, that is, as an existence of purely corporeal being, but as a "product" of the psychophysical possibilities of reception of our ego and its aprioristic forms of intuition (Kant). All the existential facts of lived experience thus acquire an altered meaning for the soul, and a new position and meaning in relation to one another. Existence enters into a new "aggregation" for the soul without assuming a new external aggregation. Though usually less intelligible and far-reaching, the same holds true in greater or lesser degree for all other spiritual expositions of the civilizational cosmos.

Now, the "soul" in each historical body attempts to form the stuff of existence—which becomes, as we have seen, the stuff of lived experience—according to its own essence, to make it the expression of its *inner being.* In this way it creates "culture." It makes no difference, therefore, whether it does this through the transformation of its corporeal existence or through a new spiritual picture of existence. Set forth through this kind of "new aggregation of all its elements of life," it sees itself, in either case, being forced into a new existence, a new world with new stuff that it has to form. In every such situation, its task begins anew. *And here originates the drive and the necessity of its "cultural productivity."* This productivity is nothing other than the attempt to give this new existence, this differently stratified stuff of life, a psychic formation.

It follows that periods of cultural productivity will always be the result of new aggregations of the elements of life. Conversely, when this "new existence" is psychically formed or expressed, there follows, necessarily, a *cultural stagnation*, perhaps relieved for a time by stilted reiterations of things already expressed, but, ultimately, utter quiescence. From the standpoint of the *"collective soul"* of a historical body (scientifically expressed, its total psychic constitution at any point in time), this means that with the new aggregation, a new *"life-feeling,"* a new way of feeling life as a *collective actuality*, arises and struggles toward expression, toward a new configuration, a new *psychic* configuration, toward a new total position with regard to social and spiritual facts. New "ages" and cultural periods arise with new sensibilities. From the standpoint of the "productive spirits" of the different historical bodies, it means that *these spirits* form the new sensibility and give rise to it in the objective world. They absorb the reaggregated stuff of life as lived experience, unify it with their psychic center, transmute it on the hearth of their new life-feeling and place this new birth, their "creation," into a "synthesis of personality and world." They do this with the intention either to simply produce a purposeless likeness of the newly felt world and its psychically formed content in a self-contained gestalt—thus making a work of art—or to psychically transform the natural, social, and civilizational form of this existence itself, to make it the expression of the new life-feeling and its content, to reconfigure and recast it in *"ideal"* form. Or, if this is impossible, if the existence that corresponds to this ideal formation appears to be inaccessible and in this sense without value, they attempt to "rescue" the emanation of the psychic from existence, by placing it beyond existence, that is, by guiding it into the transcendental. In this way, systems of ideas and religion, characterized by an orientation toward both this world and the next, arise next to works of art. And the great artists, the prophets, and the revealers of the new life-feeling arise, who embody this variegated "emanation," and start, conclude, or "crown" the epochs and periods of culture.

Thus, a very preliminary examination of the following things makes their sociological aspect and significance comprehensible in a very much superficial way: the rhythm of cultural movements, the alternation of their periods of productivity and stagnation, the advent of their "ages," the struggle of their cultural currents (which is always a struggle for expression between one such new life-feeling and another that is older), the rise of great men (who, accordingly, to a certain extent, must generally stand in "places of rupture" in the development), and the inevitable flocking together of lesser productive spirits around the appearance of the greater (the lesser "seeking" expression and pointing to the greater and the unique as "harbingers" or "allies"). The sociological type of the cultural movement, its articulation in ever new and separate periods of productivity, the conflicting currents of culture and the position, within these, of great men somewhat like standard-bearers, the protuberance-like character of great cultural emanations possessing eternal content and always characterized by complete exclusiveness and uniqueness, emanations which the movement of culture places in such polar opposition to the development of civilization—all these become comprehensible. At the same time, as intimated, the general *direction* and *form* of expression have already become more or less clear: the fact that the great cultural emanations can, at one time, signify a retreat from life (as with many of the great religious systems, Buddhism and early Christianity, for example, which despair of ever extending the psychic emanation to life as a whole); that at another time they believe themselves able to transform this unaggregated existence by means of secular idealism (Mohammedanism, Lutheranism, German Idealism); whereas at a third time they are able to accept life joyfully as it is and, in the forms of an affirmative life-feeling, simply bring it to expression in a heightened manner and complete form (Periclean Antiquity, the late Renaissance, for example).

It will be the task of sociological research to exposit these modes of *fragmentary or complete life-feeling* and their striving for expression in different forms and conditions, to connect them with spiritually or corporeally created reaggregations of the elements of life, and in this way to explain—or more cautiously expressed—to interpret not only the *great periods* of cultural productivity, their repetition, their essential nature, and the place of great men within them, but also the emergence of the *different aspects* of cultural expression, the succession and alteration of their formal principles.

The following can be offered here as a supplement with regard to the principles of cultural dynamics: each period of culture that follows from some new life-feeling, since it seeks to shape the stuff of existence and its social and civilizational aggregation and to lend it its own psychic face, naturally has an effect *back* upon this corporeal and civilizational aggregation. It creates principles of formation that are conserved and propagated in religions by the church and in systems of ideas by the spirit and the idea. It creates eternal images of objective formation in works of art, and uses great men to personify "models" of the configuration of life. Through social and spiritual channels, it impels all these into all the pores of the social and personal configuration and over the whole corporeal and spiritual habitus of the historical circle in which it arose. In this way, it sinks down, with its principles of formation, into the *subsoil* of the social and civilizational stream of historical development and, there, saturates it. This is exactly its task and purpose as the psychic form of expression of the new aggregation of life. It thus influences most intensively the course of social development and the civilizational process in every historical body. *Its* further development from the natural forces of impulse, will, and intellect is therefore complex—indeed, it is almost always in conflict with cultural formation arising from the decline of the *previous* aggregation of life. (One can recall as a historical example the self-assertion of the early capitalistic aggregation, which was a gigantic naturalism of will in conflict with the medieval, which was a *psychically* and culturally formed and fixated life.) Culturally acquired formation and rigidity can, in fact, bring the process of life-aggregation to a standstill at a certain stage by the establishment of rituals and the constraint of all natural forces (India's religiously fixed caste system). Through the contents of representation constrained in this way by ritual, it can encapsulate the civilizational process. And through all of that, the cultural formation itself becomes, reciprocally, an essential element in the concrete configuration of the society and civilization of the different historical bodies. But this does *not* alter the fact that the developments of these processes—insofar as the *social process* is borne by the natural forces of drives and will and the *civilizational process* by the intellectual forces directed toward the mastery of existence—are *primary*. Through each new life-aggregation created in this way, they present the cultural movement and, its inmost center, the soul, at any given moment with new situations and new tasks. Only then does *their* concrete *solution* of these situations and tasks create the forms and rigidities in which historical bodies are from time to time arrested, the forms from which their natural and intellectual forces continually try to liberate them. The result is the creation of ever new psychic situations, a new ground for cultural productivity. Social process, civilizational process, and cultural movement stand in such a *correlative*, reciprocal, dynamic context, whose concrete character for each historical body and for each historical instant must be elucidated by further monographic study, though in principle it must always follow the schema developed here in broad outline.

In addition, we have historical bodies and periods in which the cultural movement—for reasons yet to be disclosed by more detailed monographic studies—reacts less powerfully upon the "natural" formation of the whole body or upon certain distinct configurations than at other times; periods in which cultural movement allows the natural configurations to evolve more or less according to their own laws and gives expressions to the envisioned life only in a psychically created form. Such a period, for example, is Classical Periclean Antiquity, where economy, family, even the state to a certain degree, and, on the other side, "cognition" could live and develop entirely according to their own natural laws; and only the state as "polis," as a closer community of people, was given a religious foundation. Works of art and ideas brought life formed in this manner into connection with ultimate psychic judgments and accordingly gave it its expression. Near Eastern-Egyptian Antiquity represents an entirely opposite type. Here, at least, the state and cognition were completely enclosed by religious forms and were brought into uniformity and into a culturally fixed expression of a ritualization, which gave support and strength to the outcome of this peculiar social formation (bureaucracy!). The same may be said of the Indian or Chinese historical circles, though in quite different fashion and on quite different social grounds. At times, the medieval feudal aggregation of life manifested tendencies of a similar rigidity of life under the influence of the church. The principal characteristic of the Renaissance was that the natural-social and the civilizational-rationalistic forces burst those bonds, set the social and civilizational processes free again, and gave the state, economy, and cognition a natural "configuration" by "undeifying" them so that—in real, though remote and exaggerated resemblance to Antiquity—"culture" could join into a closer relationship with natural life, as it had in the ancient world. The main point at issue is this: the various social structures, economy, state, class, family, and so on, as well as the spheres of civilization and their parts, could attain different degrees of "cultural saturation" in the various historical bodies and in various periods, and could, therefore, vary widely in the degree to which they were bound in a definite form. This is always a function of the "correlation" of the three spheres, the interconnections of which still have to be examined more closely.

Perhaps it would be pertinent to our *current position* and our present relationship with *it* to make ourselves clear on the following: the *psychic* elaboration of the "modern" aggregation of life (which emerged out of the Renaissance and of which there was evidence in the early capitalistic period) really began only in the Idealist period of the eighteenth century. Only *then* was there a necessity to bring a new psychic-cultural form to the new civilizational and natural forces that the "modern" configuration of life brought forth. But this was not accomplished. The tremendous power of the natural social and civilizational forces, which have been released since the Renaissance and in whose *initial period* of impact the modern cultural forms arose (although it was believed these forces could again be contained), exploded all these plans and gave rise to the nineteenth century with its upheaval. But all the "cultural postulates" we possess today to cope with the natural forces and tendencies of our life, *all* that we possess to cope with the domains of economy, state, society, and family in truth still stem from the psychic-spiritual arsenal of that earlier attempt at mastery. We have not yet created a new ideology with which to confront our *present* aggregation of life. Socialism in *all* its present ideational forms is really the postulate of a new cultural form of economy and society directed by that *early* life-feeling against capitalist forces, a postulate laden with all the frailties and anachronisms of that former psychic-spiritual attitude. The modern theory of the state is in the same quandary, and no other has as yet been thought out to reckon with and culturally master the *actual* new aggregation of life that has since taken place. The new "theory of the family," the center of so much conflict today, is having the same difficulties in its sphere. And so forth to the idea of "nation" and of "humanity," these last-to-be-conceived and highest organizations of humanity in its totality, which we still try to grasp and conceptually order in terms of the tentative schema of the eighteenth century.

Today, we continue to fight the battle of the modern soul and the modern aggregation of life. We fight with the *old*, half-blunted weapons of this earlier time without the possibility, as yet, of a new psychic grasp of the totality of life that would put new, more effective weapons into our hands. Therefore, our battle is harder and more desperate than ever. And due to the gradual and continued regression from higher aims, today these *most essential* objects have already become the most simple of all basic social configurations in the natural foundations of existence. If everything has been rooted up—according to the specific nature of the aggregation of life— these foundations can easily become the real and primary object of the pressure of psychic demands, a pressure which then must, so to speak, submerge and attempt to orient itself in a radically new way down there at the roots of existence.

This much, however, can be said in general conclusion: cultural movement has widely varying degrees of success in drawing the social and civilizational products into its path, depending on the time and the historical body. Moreover, its *will* to do so varies just as widely in different times and periods; for the life-feeling of the soul, which is confronted by a certain aggregation of life and grows out of it, is only *capable* of this effect within the limits of its own strength. It sees in varying degrees the possibility, or, in happier "times," the *necessity*, of a complete formation of the stuff of lived experience. All this has to be the object of a closer study of cultural sociology. Here, it is chiefly a matter of the clear presentation of the fundamental nature of the concepts of culture, cultural movement, civilizational and social processes, and their reciprocal dynamic relationship, in so far as it is possible in brief indications.

VII

From this standpoint, one can take a position with respect to the two ways, historico-philosophical and sociological, in which cultural movement has generally been considered, namely, the "evolutionary" approach and the one recently termed "morphological."

The evolutionary, historico-philosophical approach to cultural movement has its origin in the confusion of the intellectual and psychic spheres under the collective concept of "spirit" and, consequently, in the inclusion of civilizational process and cultural movement under the collective idea of "spiritual development," a blending for which the eighteenth century paved the way and which German Idealism brought to its climax. As a consequence, the civilizational process and cultural movement are so entangled that the nomononies of *civilizational* development, first in the guise of "spiritual development," and then in that of general "development," have been paraded as the form of perception for the historico-philosophical study of the total process of human history. Condorcet regards all history as the process through stages of man's perfection, the content of which is, in truth, "enlightenment," that is, in our own parlance, entirely and exclusively the exposition of a part of the civilizational cosmos. Kant, Fichte, and Hegel, despite the difference in their sociological constructions, were agreed in viewing the content of history as the elucidation of consciousness (another aspect of the civilizational process) with the goal of the exposition of the "consciousness of freedom," which should establish the realm of reason. It is immaterial whether this is revealed in more rational forms (attraction and repulsion, solidaristic and individualistic forces, whose interaction ultimately leads to the realm of reason—Kant), whether it receives a Biblical-Protestant cloak as in Fichte (stage of innocence, of growing and complete sinfulness, of liberation, and finally of the supremacy of reason), or whether it assumes the grand Hegelian form of evolution of the world spirit that utilizes the drives and passions of humanity and its struggle for clarity and for rational order, so that through thesis, antithesis, and synthesis, that is, the unfolding in a *logical* process of self-development, it finally arrives at the establishment of the realm of "reason" incarnated in the state. It is clear that the development takes place, in each case, according to logical and intellectual principles. And the finally exposed realm of reason into which the individual must incorporate himself in "freedom of consciousness," is at the bottom none other than our elucidated and exposed civilizational cosmos, which absorbs everything else (art, religion, ideas, etc., and all cultural emanations) as elements of its "rational" progress, engulfing them all in its final, rational form. The view of the essence of the *civilizational process*, which was first perceived psychologically, obscures the view of the essence of culture, and, under the concepts of the realization of reason and the intrinsic evolution of spirit, draws the totality of historical facts into its representational province. Marx is fundamentally caught up in the same situation. What he sees is simply another side, the *material and technical* side of the unfolding of the civilizational cosmos, and lets *it* appear as the original principle of historical development, of which all social processes are mere expressions and all cultural movements, mere reflections. The positivists, from their gifted founders, St. Simon and Comte, to the present pragmatists, perceive the line of spiritual and *scientific* development, see the ever more rapid removal of mythical veils from the world picture, see the constantly growing influence of intellect and science on the structure of existence and society (industrial system in St. Simon), and finally allow everything, including culture, which naturally they do not recognize as being of different essence, to vanish in their civilizational and positivistic pattern of world and existence—which, again, is equivalent to the exposited cosmos of civilization. Therefore, as far as the manner of observation is concerned, it is immaterial if they then attempt to give this rationalized and thoroughly organized civilizational cosmos (exposited as the final goal of culture) a religious consecration appropriated from accentuation of values of the cultural (as did St. Simon in his *Nouveau christianisme*). Later sociologists like Spencer, who are under the spell of the positivistic civilizational attitude, organize the resulting perception of the whole historical happening, not according to the developmental facts of an objective spirit, the technical means of production, or the scientific mastery of existence, but according to the reflection of all these things back on the development of consciousness itself. Thus the militant, myth-making, and religious man of the early stages is succeeded by the rationalized, mercantile, and compassionate man as a necessary product of development. Or one views the effect of the development on the conscious attitude of the individual toward the collectivity. The early periods of complete corporative bondedness are succeeded by a period of individualism and, perhaps, finally by a period of so-called subjectivism that has never been clearly defined (Lamprecht). In all cases, it is simply the progress of civilization and its effects that are being viewed and investigated. Everything else is merely part, consequence, or reflection of this progress. Everywhere, the exposition of some aspect of the preexistent civilizational cosmos, along with its effects, is viewed as the content, purpose, and goal of world history. For the consciousness of all these people, this civilizational cosmos, in its latest part (whose emergence they are able to behold), is something definitive, the final goal toward which we must strive. And hence, it always happens that evolutionary civilizational sociologists and philosophers of history, even the most gifted among them, nay, particularly those, all offer eschatologies, predictions, or confirmations of a final stage of humanity, a final stage that is in all cases simply the last stage they have beheld in the unfolding of the civilizational cosmos. As is well known, Hegel and Fichte considered their age as that of the dawning of practical reason, as the last age of men—a slight error, as we know today. Marxism, with its predictions of the socialist future realm of rationalism, which would finally emerge in a logical and dialectical manner through a social process viewed as purely civilizational, is just such a civilizationally conceived eschatology, the attempt at a preliminary elucidation of the future civilizational cosmos and its forms to sharpen social and political agitation.

It is not strange in the least that all these various theories of history and philosophies of culture, different as they may be in their self-proclaimed principles—idealist, materialist, positivist, psychological, and so on—are nevertheless so fundamentally contiguous that, on closer scrutiny, one unexpectedly merges into the other; in fact, one is nothing but the reverse side of the other. For example, this has been convincingly demonstrated by Plenge in the relationship (not simply in form but also in content) between, indeed, the far-reaching sociological identity of Hegelianism and Marxism, whose philosophers of history confronted one another so inimically in their outward demeanor (evolution of spirit versus evolution of matter), while at bottom both asserted the necessary and universally valid intrinsic evolution of a rational, social collective organization of humanity, in whose necessity the individual had only the freedom "to adapt" in his consciousness.[13] But this same far-reaching identity can be shown in the whole complex of all these theories of development and culture. For, they are all merely different illustrations of the collective rational principles that form and develop the inner and outer aspects of human existence; and these principles are allowed to obscure everything else.

All these authors stand, as it were, sociologically *prior* to the "original sin" of basic insight. They do not see that rational organization, rational self-incorporation, rational elucidation of existence, or any other rationalization—even if we project into them the emanation of "world reason" or relate them to the development of principles of freedom or to trends toward equality, even in this ideological clothing—still have nothing *as such* to do with either culture or especially the formation of a historical body from the psychic center of its essence. They are simply not aware of the distinction between the civilizational tendencies of development in the historical bodies and the unfurling of culture. Otherwise they would perceive that all these processes of rationalization are only *means* and not the *essence* of the configuration of existence. They would not be able to see any goals or ultimate ideas of human development in these facts, but rather the developmental generalities and necessities *under whose* widening mastery the psychic is located in the historical process, and take as their task the unending and ever harder struggle to conquer the natural existence in which generalities and necessities themselves have been among the creative factors. Of course, the eschatological, civilizational paradise of organization that they all see would then be deprived of divine attribute, and humanity of its general "goal of culture." But with that, they would have gained something most profound and ultimate concerning the question of culture proper.

VIII

The "morphological" approach to history and culture is the direct antithesis. It seeks to understand the "soul" as it matures, awakens, lives its life and becomes old in the various great historical bodies, and as it exposits cultural emanations as symbols of its existence and destiny. The religions, systems of ideas, and configurations of art are the utterly unique, incomparable forms of expression of the soul of the various great historical bodies in their different stages of blossoming and becoming old, forms that strive toward no universal human goal but have simply become "gestalt." History itself is not an inherently unified coherent process, but the province of morphological emergence, of the origin, growth, and disappearance of these great bodies from a "maternal landscape." Each historical body has its own law, its own essence, and its own peculiar soul that strives for expression. They all have "homologous" stages of development and "homologous" drives for expression, to the extent that they are all young at some time, that they all grow up, bloom, and become old, that they all seek to reflect the totality of their psychic content in their forms of manifestation. But each single aspect of manifestation, civilizational as well as cultural, is as unique, exclusive, and incomparable as its soul itself, a pure expression of its essence. Thus, we have not only a Faustian-Occidental, an Apollonian-Classical, an Arabian-Magian art, religion, and metaphysics, but a science and mathematics as well, a particularly configured, nontransferable system of knowledge—in our parlance, an inherently nontransferable civilizational cosmos peculiar to each historical body, especially appertaining to its essence, and neither universally valid nor necessary. All parts of the civilizational cosmos are viewed, in general, not as civilizational but as "cultural," that is, as psychic forms of expression. Civilization and culture are thus once again fused into one, as with the evolutionists, but this time from the opposite side. Culture and civilization are distinguished from one another only to the extent that the civilizational formation is viewed and defined as the conscious, rational, "cosmopolitan" final form of the historical process of each great body, whereas culture, at a definite "homologous" stage of aging in its development, is said to devolve into civilization, that is, "rational senility." Thus, the essence of the civilization is seen and recognized to a certain extent as a rational formation of life and as conscious elucidation of existence. But the civilizational process proper is not *recognized* as a *unified process* permeated by the destiny of mankind, a process present from the very beginning of all historical development and permeating all the stages peculiar to each historical body at all times. Undoubtedly, the increasing significance of civilizational formation, which follows from the elucidation of consciousness, is correctly seen. This elucidation of consciousness, however, is not placed in its proper context, namely, that of the continual repetition and advancement of the *general* elucidation of human consciousness that was begun or already realized in *other* bodies. Even for this view, humanity's *objective* civilizational cosmos of this elucidation must be presented as a clearly evident and undeniable whole, hardly fragmentary even in earlier times of less universal connectedness. Furthermore, those facts of the course of history, which actually have their own laws, that is, the realities and forms of manifestation of the *social process* of the various historical bodies, are interpreted arbitrarily as symbols of the will for psychic expression, bypassing each instance of causality. In short, not only the civilizational but also the social process is drawn into the "morphologically" viewed movement of culture in order to arrive at the respective intrinsic process of growth, aging, and final destiny of the various great historical bodies. This presents a picture that has many laughable aspects. For instance, it is laughable to predict the aging of Occidental culture in the future, giving it a hundred years' grace up to such and such a homologous stage. And this at a moment when this Occidental culture is caught up in a gigantic process of transformation through its involvement in the general destiny of mankind (the World War!), a process which is propelling culture—whether toward the disintegration of its historical body and its "soul," whether toward metamorphosis or transposition into another emergent body, or toward a complete and perhaps very rapid "physical" decline—toward a completely unknown and absolutely incomprehensible destiny. The fallacy in this view follows from its characteristic inability to separate social process, civilizational process, and cultural movement, and from its attempt to understand and define the destiny of the various historical bodies by giving sole validity to some comprehended adherence to the soul, without seeing and investigating the realities of the social and civilizational processes and their specific nomonomies.

IX

Nevertheless, there can be no doubt that our way of conceptualizing *culture* is akin to the morphological approach. For us, too, all cultural emanations are merely unique symbols, pointing to no general goal of development—mutually incomparable and exclusive structures of expression of something psychic, the soul of the existing historical bodies that propels them into the world. They are, however, symbols that receive their *form* and *content* of expression essentially from the extant aggregated life-substance, from the social and civilizational corporeality in which this "soul" appears in the various historical bodies at various times, and which the soul, through this symbolization and exposition in the form of culture, attempts to make into its own "body" (*Leib*). The material of society and civilization surrounds the soul of the various historical bodies with a stuff alien to the soul, a material layer to which, at each historical moment, the soul seeks to lend its own "countenance," or, failing in this, to withdraw in abnegation (the great renunciatory religions). Or, if one prefers another metaphor, a stuff, a life-material obtrudes itself upon the soul according to its own inevitable laws and demands that the soul inspire it and give it a cultural form. This is the cultural *task* of each era and the *essence* of its cultural movement, which, to repeat, is still therefore primarily dependent on factors *other* than the "intrinsic development of the soul of the historical bodies." Even in this view, there is room for an intrinsic unfolding of the soul, a growth, a bloom, and a process of aging. All this, however, cannot be considered as self-sufficient and independent of causality. It is conditioned by the elucidation of consciousness of the different peoples' life-substance that is created by the civilizational process; it is conditioned by the successive elucidation of these peoples as they, one after another, enter into the subjective aspect of the civilizational process of humanity. This entrance allows the peoples of the different historical bodies to have periods of youth, which lack awareness and enlightenment and which vary greatly in both type and duration, depending on the time, manner, and place of their entrance into this cosmos. This entrance allots, to the various historical bodies, periods of psychic awakening under widely varying conditions in widely varying settings of psychic and physical objects already offered up by the civilizational cosmos. And it leads the historical bodies, in extremely diverse ways, to high periods of conscious, productively forming, psychic domination of the collective existence; and it allows an aging that is in no way universally similar or predetermined. Further, just as the whole unfolding of consciousness and the process of psychic growth is conditioned by the time, place, and manner of entrance into the general human *civilizational cosmos* and is thereby conditioned in tempo and in the emergence of its initial and middle stages, so is it likewise conditioned in its further development (as our cultural sociological investigation of the various historical bodies will show) by the widely varying course and structure of the *social process*. This latter condition may take several forms: completely unconscious or half conscious strata of the historical body in question may at first lag behind and, only with its progress, ascend and realize anew in themselves the development of consciousness (such as the succession of ascending strata of clerics, knights, burghers, workers, etc. in the Occident). Or, a simultaneous general enlightenment of a people (from a social perspective) may occur as in Classical Antiquity (a development in one stage!). Or, a fixed social and spiritual organization of hierarchically arranged strata may result from the dynamics of the social, civilizational, and cultural process as in India and other historical bodies. In our view, therefore, the process of psychic elucidation, growth, and becoming old of the cultural body, together with all its cultural expressions and possibilities, is embedded, like everything else, in the mutual dynamics of the processes of society, civilization, and culture. It, accordingly, has entirely different points of departure in the different historical bodies. (The Arabian historical circle began at a stage of consciousness entirely different from the Classical because of its different position in the civilizational cosmos; and it never knew a real mythology.) The process also has quite varied possibilities of development (single stage, multiple stages of development, etc.) And by means of the varied interweaving of the historical bodies in the total process of history—from remaining in more or less complete isolation (as in China or India) to interaction and even ultimately world-wide expansion (as in Classical Antiquity, Arabia, and the modern Occident)—it flows into completely different final states not amenable to schematization. All this, the phases and forms of expression of cultural "movement" as well as the "destiny" of culture in the different bodies, is to be explained through a treatment of facts within the framework of a perspective that never loses sight of the *total* course of human historical development that must operate, as has already been explained, on three planes and must carefully separate three parts.

First, *something* must be accepted simply with no attempt to analyze or explain it—namely, the *specific* quality of the soul that "dwells" in the various great historical bodies and constantly struggles for expression within the compass of their general destiny. This specific quality of the soul is to be comprehended exclusively through "empathy." For anyone who understands this empathy, it is possible to grasp the psychic factor in every cultural objectification, interpret it in its essence, interpret and lay it bare as the specific "core" of the culture in question. But the sociologist, at least with respect to the aims pursued *here*, must leave this untouched. Its interpretation and presentation must be left in more delicate hands—as well as the interpretation and presentation of the deepest meaning and essence of all great cultural objectifications in general.

He should be primarily concerned with a second, antithetical thing, namely, the "surface movement" that this "core" brings forth in its struggle with the external stuff of life. In this surface movement, he must establish the typical and the recurrent; he must establish the succession of emanations, the detachment of aspects and forms of expression, the typical periodicity, the manner and kind of impact—in other words, the "rhythm of cultural movement"; and he must do this, of course, as far as possible in all historical and cultural bodies that lend themselves to analysis. In this analysis, he must continually ask to what extent the established typology coheres with the corporeal and the intellectual-civilizational (with the processes of society and civilization) that surround the psychic "core," that is, to what extent this typology follows from the situations, modes of life aggregation, in which these place the psychic core. In this connection, the sociologist must present inductively a general *formal typology* of cultural movement, clarifying sociologically the surface movement of cultural development. Such a study takes its place side by side with the dynamics of the different spheres, as we have propounded them.

Third, he can attempt to go deeper and inquire into the special *destiny* of the "psychic core" in this movement. He can follow the psychic core of the various historical bodies in its *development* within the social and civilizational processes. He can examine the "changes" *in content* by which the core "struggles" with its "destiny," a destiny imposed upon it by the processes of society and civilization. He can perhaps throw light upon how the advent of great men follows from this struggle and these "changes," how the great "ages," cultural currents, and countercurrents result, how the great *material* lines of cultural history become evident. Perhaps! He should at least make the attempt. He would thus fill the *formal* sociological surface typology of cultural movement with a *material content* that could be understood sociologically and, at the same time, create a bridge to the interpretation of *contents* and apprehension (through understanding) of the *essence* of the great cultural objectifications and phenomena, which (as we have said) it is *not* his task to interpret in their deepest meaning. But to understand them adequately, he can build an environment after *his own* fashion, a sociological rigging and shells, as it were, in which the golden balls of cultural history lie. In this way, perhaps, they can be seen to greater advantage and their essence more easily grasped than through mere "empathy" in "empty space." In this last attempt, he will have almost overstepped the bounds of the pure sociological approach. But if he gets this far, he will make his contribution to our ultimate goal in the discerning understanding of cultures: their psychic renaissance in ourselves.

Further essays will take up this task with the limited means that are available to the author—but certainly without the ambition to deliver definitive results on any point.

Appendix 3: Cited Works by Alfred Weber

A. *Alfred Weber-Gesamtausgabe*, eds. Richard Bräu, Eberhard Demm, Hans G. Nutzinger, Walter Witzenmann. Marburg: Metropolis, 1997–2003. [AWG]
Vol. 1: *Kulturgeschichte als Kultursoziologie*, ed. Eberhard Demm. 1997.
Vol. 2: *Das Tragische und die Geschichte*, ed. Richard Bräu. 1998.
Vol. 3: *Abschied von der bisherigen Geschichte / Der dritte oder der vierte Mensch*, ed. Richard Bräu. 1997.
Vol. 5: *Schriften zur Wirtschafts- und Sozialpolitik, 1897–1932*, ed. Hans G. Nutzinger. 2000.
Vol. 4: *Einführung in die Soziologie*, ed. Hans G. Nutzinger. 1997.
Vol. 6: *Schriften zur Industriellen Standortlehre*, ed. Hans G. Nutzinger. 1998.
Vol. 7: *Politische Theorie und Tagespolitik, 1903–1933*, ed. Eberhard Demm. 1999.
Vol. 8: *Schriften zur Kultur- und Geschichtssoziologie, 1906–1958*, ed. Richard Bräu.
Vol. 9: *Politik in Nachkriegsdeutschland*, ed. Eberhard Demm. 2001
Vol. 10: *Ausgewählter Briefwechsel*, ed. Eberhard Demm & Hartmut Soell. 2003.
B. Individual works with AWG reference.
1897. "Hausindustrielle Geseztgebung und Sweating-System in der Konfektionsindustrie," [*Schmollers*] *Jahrbuch für Gesetzgebung, Verwaltung und Volkswirtschaft* 21: 271–305 [AWG 5: 25–58].
1897. "Neuere Schriften über die Konfektionsindustrie," *Archiv für soziale Gesetzgebung und Statistik* 11: 527–40 [AWG 5: 83–96].
1897. "Die Entwickelung der deutschen Arbeiterschutzgesetzgebung seit 1890," *Schmollers Jahrbuch* 21: 1145–94 [AWG 5: 97–140].
1897. "Das Sweatingsystem in der Konfektion und die Verschläge der Kommission für Arbeiterstatistik," *Archiv für soziale Gesetzgebung und Statistik* 10: 493–518 [AWG 5: 59–82].
1899. "Die Entwicklungsgrundlagen der großstätischen Frauenhausindustrie," *Schriften des Vereins für Sozialpolitik* 85: xiv-lx [AWG 5: 175–221].
1899. "Der Arbeiterschutz in der Konfektion und verwandten Gewerben," *Soziale Praxis; Centralblatt für Sozialpolitik* 8: 689–94 [AWG 5: 146–52].
1899. "Beschränkung der Heimarbeit in der Konfektionsindustrie," *Soziale Praxis; Centralblatt für Sozialpolitik* 27: 721–27 [AWG 5: 151–62].
1900. "Die Hausindustrie und ihre gesetzliche Regelung: Referat," *Schriften des Vereins für Sozialpolitik* 88: 12–35 [AWG 5: 222–45].
1901. "Die volkswirtschaftliche Aufgabe der Hausindustrie; Akademische Antrittsvorlesung," *Schmollers Jahrbuch* 25: 383–405 [AWG 5: 246–66].
1902. "Die Arbeitslosigkeit und die Krisen," *Die Frau* 9: 449–54, 548–54 [AWG 5: 319–34].

1902. "Deutschland am Scheidewege," *Schmollers Jahrbuch* 26: 1293–305 [AWG 5: 335–45].
1903. "Deutschland am Scheidewege; Eine Replik," *Schmollers Jahrbuch* 27: 275–89 [AWG 5: 346–59].
1904. "Deutschland und der wirtschaftliche Imperialismus," *Preußische Jahrbücher* 116: 298–324 [AWG 5: 382–406].
1906. "Theodor Mommsen," *Die neue Gesellschaft* 12 [AWG 8: 83–90]
1907. "Konstitutionelle oder parlamentarische Regierung in Deutschland?" *Neue freie Presse* 15324 (Apr. 21) [AWG 7: 32–41].
1909. *Über den Standort der Industrien: 1, Reine Theorie des Standorts.* Tübingen [AWG 6: 29–265].
1909. "Der Kulturtypus und seine Wandlung," *Heidelberger akademischer Almanach*: 51–61 [AWG 8: 91–97].
1910. "Der Beamte," *Die neue Rundschau* 21: 1321–39 [AWG 8: 98–117].
1910. Contribution to discussion of "Die wirtschaftlichen Unternehmungen der Gemeinden," in *Schriften des Vereins für Sozialpolitik* 132: 238–48, 309–12 [AWG 5: 425–36].
1911. "Die Standortslehre und die Handelspolitik," *Archiv für Sozialwissenschaft und Sozialpolitik* 32: 667–88 [AWG 6: 267–85].
1911. "Der Mann von 40 Jahren." Unpublished Manuscript. Bundesarchiv Koblenz, NL 197 Alfred Weber 8. Pp. 51–52 in Demm (ed.), *Alfred Weber zum Gedächtnis.* Frankfurt/Main, 2000.
1911. "Der produktive Geist," *Neue freie Presse* 16685 (Feb. 2) [AWG 8: 254–59].
1912. *Religion und Kultur.* Jena [AWG 8: 315–38].
1912. "Das Berufsschichsal der Industriearbeiter," *Archiv für Sozialwissenschaft und Sozialpolitik* 34: 377–405 [AWG 8: 344–68].
1912. Untitled contribution. Pp. 127–33 in Werner Sombart et al., *Judentaufen.* Munich [AWG 8: 339–43].
1912. Letter to Hermann Keyserling, Dec. 2, 1912. Bundesarchiv Koblenz, NL 197 Alfred Weber 49 [AWG 10: 500–2].
1913. "Der soziologische Kulturbegriff," *Schriften der deutschen Gesellschaft für Soziologie* 2: 1–20, 74 [AWG 8: 60–75].
1913. "Neuorientierung in der Sozialpolitik?" *Archiv für Sozialwissenschaft und Sozialpolitik* 36: 1–13 [AWG 5: 475–85].
1913. "Die Bureaukratisierung und die gelbe Arbeiterbewegung," *Archiv für Sozialwissenschaft und Sozialpolitik* 37: 361–79 [AWG 5: 459–74].
1913. "Geleitwort." Pp. i-vi in Staudinger 1913 [AWG 8: 373–78].
1913. "Leben und Kultur; Beiträge zur Erkenntnis ihrer inneren Beziehung." Unpublished typescript. NL 197 Alfred Weber 14. [Draft of 1913f.]
1913. "Program einer Sammlung 'Schriften zur Soziologie der Kultur,'" *Die Tat* 5: 855–56 [AWG 8: 379–80].
1913. "Schule und Jugendkultur," *Frankfurter Zeitung* 248, Sept. 7 [AWG 7: 83–90].
1914. "Individuum und Gemeinschaft," *Die Tat*, 5: 887–900 [AWG 8: 381–83].
1914. "Rede über die freideutsche Jugendbewegung auf der Aufklärungsversammlung in der Münchener Tonhalle am 9. Febr. 1914 nebst Schlußwort." Pp 11–19, 43–45 in *Die Freideutsche Jugend im Bayrischen Landtag.* Hamburg [AWG 7: 92–105].

1914. "Eine Zuschrift," *Die freie Schulgemeinde* 4: 45-47 [AWG 7: 106-8].
1914. "Industrielle Standortslehre (Allgemeine und kapitalistische Theorie des Standortes)," *Grundriss der Sozialökonomik* 2, new pt. (1923) 6: 58-86 [AWG 6: 287-328].
1915. *Gedanken zur deutschen Sendung*. Berlin [AWG 7: 116-77].
1915. "Entgegnung [to Lukacs 1915]," *Archiv für Sozialwissenschaft und Sozialpolitik* 39: 223-26 [AWG 8: 384-87].
1915. "Bemerkungen über die auswärtige Politik und die Kriegziele." Pp. 171-77 in Demm 1986 [AWG 7: 109-15].
1917. "Kontinental Verständigung." Pp. 315-19 in Demm 1990 [AWG 7: 182-86].
1917. "Aufstellung eines Friedensprogramms mit unseren Bundesgenossen." Pp. 320-25 in Demm 1990 [AWG 7: 187-93].
1918. "Das Selbstbestimmungsrecht der Völker und der Friede," *Preussischer Jahrbücher* 171: 60-71 [AWG 7: 194-206].
1918. "Die Bedeutung der geistigen Führer in Deutschland," *Die neue Rundschau* 29: 1249-68 [AWG 7: 347-67].
1920. "Prinzipelles zur Kultursoziologie (Gesellschaftsprozeß, Zivilisationsprozeß und Kulturbewegung)," *Archiv für Sozialwissenschaft und Sozialpolitik* 47: 1-49 [AWG 8: 147-86].
1920. "Die Zukunft der deutschen Hochschulen. Geistesleben und finanzielle Lage," *Frankfurter Zeitung* 84 (Feb. 1) [AWG 7: 561-65].
1921. "Auflösung des Staatsgedankens?," *Frankfurter Zeitung* 889 (Nov. 30) [AWG 7: 368-72].
1922. "Deutschlands finanzielle Leistungsfähigkeit jetzt und künftig," *Archiv für Sozialwissenschaft und Sozialpolitik* 49: 265-97 [AWG 5: 518-45].
1922. "Vorwort." Pp. iii-viii in Otto Schlier, *Der deutsche Industriekörper seit 1860; Allgemeine Lagerung der Indistrie und Industriebezirksbildung*. Tübingen [AWG 6: 347-49].
1923. *Die Not der geistigen Arbeiter*. Munich and Leipzig [AWG 7: 601-39].
1923. *Deutschland und Europa 1848 und Heute*. Frankfurt/Main [AWG 7: 505-16].
1923. "Kultursoziologie [und Sinndeutung der Geschichte]," *Der neue Merkur* 7/1: 169-76 [AWG 8: 76-82].
1924. *Deutschland und die europäische Kulturkrise*. Berlin [AWG 7: 469-98].
1924. "Oswald Spengler der Politiker," *Der neue Merkur* 7: 773-77 [AWG 7: 448-52].
1924. "Programmentwurf für die Republikanische Partei Deutschlands." Pp. 337-40 in Demm 1990 [AWG 7: 440-43].
1924. "Die Reichstagswahlen—und nun was? Ein Brief," *Frankfurter Zeitung* 355 (May 13) [AWG 7: 446-47].
1924. "Verjüngung," *Frankfurter Zeitung* 796 (Oct. 24) [AWG 7: 453-54].
1925. *Die Krise des modernen Staatsgedankens in Europa*. Berlin and Leipzig [AWG 7: 233-346].
1925. "Schule am Meer. Offener Brief an Herrn Luserke," *Frankfurter Zeitung* 157 (Feb. 28) [AWG 7: 657-60].
1926. "Kultursoziologische Versuche: Das alte Aegypten und Babylonien," *Archiv für Sozialwissenschaft und Sozialpolitik* 55: 1-59 [AWG 8: 203-52].
1926. "Geist und Politik," *Die neue Rundschau* 37: 337-48 [AWG 7: 377-88].

1926. "Republikanische Tradition," *Der eiserne Steg Jahrbuch 1926*: 31–33 [AWG 7: 456–57].

1926. "Paneuropa," *Europäische Revue* 1: 149–53 [AWG 7: 527–32].

1926. "Rede beim Schlussbankett der internationalen Jahresversammlung in Mailand," *Europäische Revue* 1: 301–3 [AWG 7: 533–35].

1926. "Die Krise der Menschenrechte. Ist aus dem kollektiven Kulturbewußtsein eine Besserung zu erhoffen?" *Neue freie Presse* 22018 (Jan. 1) [AWG 7: 375–76].

1926. "Tschechische geschichtliche Mission und deutscher Geist. Offener Brief an den Präsidenten T. G. Masaryk," *Frankfurter Zeitung* 285 (Apr. 18) [AWG 7: 517–21].

1926. "Amerika—Europa. Zu Arthur Feilers Amerika-Buch," *Frankfurter Zeitung* 356 (May 15) [AWG 8: 392–96].

1926. "Eine Rede des Professors Alfred Weber," *Neue freie Presse* 22307 (Oct. 19) [AWG 7: 536–38].

1927. "Einleitung: Aufgaben und Methode." Pp. 1–28 in Weber, *Ideen zur Staats- und Kultursoziologie*. Karlsruhe [AWG 8: 34–117].

1927. "Der Deutsche im geistigen Europa," *Europäische Revue* 2: 276–84 [AWG 7: 539–48].

1927. "Mythologie oder Wirklichkeit?" *Europäische Revue* 3: 641–44 [AWG 7: 549–52].

1929. *Theory of the Location of Industries*. Trans. Carl J. Friedrich. Chicago. [Trans. of Weber 1909]

1929. Contribution to "Diskussion über 'Die Konkurrenz.'" Pp. 88–92 in *Verhandlungen des Sechsten Deutschen Soziologentages vom. 17. bis 19. September 1928 in Zürich*. Tübingen [AWG 8: 411–15].

1929. Contribution to "Discussion of Karl Mannheim's 'Competition'" paper at the Sixth Congress of German Sociologists (Zurich, 1928)." In *Knowledge and Politics: The Sociology of Knowledge Dispute*, ed. Volker Meja and Nico Stehr, 86–90. London, 1990.

1929. "Eigene Gedanke Alfred Webers über die Demokratie," *Frankfurter Zeitung* 768 (Oct. 15) [AWG 7: 458–61].

1930. "Zur Plattform der Deutschen Staatspartei," *Frankfurter Zeitung* 594 (Aug. 12) [AWG 7: 426–28].

1931. "Kultursoziologie." Pp. 284–94 in Alfred Vierkandt (ed.), *Handwörterbuch der Soziologie*. Stuttgart [AWG 8: 129–46].

1931. "Wie das deutsche Volk fühlt!" *Europäische Revue* 7: 89–92 [AWG 7: 522–26].

1931. *Das Ende der Demokratie? Ein Vortrag*. Berlin [AWG 7: 389–402].

1932. "Zur Krise des europäischen Menschen," *Europäische Revue* 8: 759–63 [AWG 8: 418–22].

1933. "Werner Sombart. Zum 70. Geburtstag," *Frankfurter Zeitung* 49/50 (Jan. 19) [AWG 8: 423–27].

1935. *Kulturgeschichte als Kultursoziologie*. Second ed. Munich (1950) [AWG 1].

1947. "Professor Alfred Weber, Heidelberg." Unpublished Dictation. Bundesarchiv Koblenz, NL 197 Alfred Weber, 8. Pp. 21–23 in Demm (ed.), *Alfred Weber zum Gedächtnis*.

1946. "Zu dem warnungen F.W. Forsters in der deutschen Frage," *Neue Züricher Zeitung* 1961 (Oct. 30) [AWG 9: 71–75].

1948. *Farewell to European History, or The Conquest of Nihilism*. Trans. R. F. C. Hull. New Haven, 1948.
1949. "Haben wir Deutschen seit 1945 versagt?" *Die Wandlung* 4: 735–47 [AWG 9: 91–102].
1955. *Einführung in die Soziologie*. Munich [AWG 4].
1955. "Die Jugend und das deutsche Schicksal: Persönliche Rückblicke und Ausblicke." Pp. 56–73 in Elga Kern (ed.), *Wegweiser in der Zeitwende*. Munich/Basel. [AWG 8: 617–33.]
1960. "Soziologie." Pp. 595–632 in *Propyläen Weltgeschichte*. Vol. 9. Berlin. [AWG 8: 709–57].

Notes

Introduction

1. H. Stuart Hughes, *Consciousness and Society* (New York: Vintage, 1961), 17. Hughes attributes the term to A.L. Kroeber.
2. None of the scholars I will be discussing, especially in the early parts of this work, would have considered himself a "sociologist" during the empire. Most of them, including Weber, thought of themselves as national economists.
3. Wilhelm Hennis, *Max Weber: Essays in Reconstruction*, trans. Keith Tribe (London: Allen & Unwin, 1988), 22. More than two decades before Hennis, Friedrich H. Tenbruck had lodged a similar complaint. Tenbruck, "Die Genesis der Methodologie Max Webers," *Kölner Zeitschrift für Soziologie und Sozialpsychologie*, 11 (1959), 573–630.
4. The importance of Fritz K. Ringer's *The Decline of the German Mandarins* (Cambridge, MA: Harvard University Press, 1969) must be acknowledged.
5. Wolfgang Schluchter, "Max und Alfred Weber—zwei ungleiche Brüder," 1–7, http://www.uni-heidelberg.de/uni/presse/rc7/5.html (last accessed April 10, 2012). Christian Jansen notes that from early on Alfred was referred to as "mini-Max." Jansen, "Neues von der deutschen Gelehrtenpolitik," *Neue politische Literatur* 51 (2006), 17.
6. Bernd Widdig, *Culture and Inflation in Weimar Germany* (Berkeley & Los Angeles: University of California Press, 2001), 183.
7. Robert Michels, *Political Parties: A Sociological Study of the Oligarchical Tendencies of Modern Democracy*, trans. Eden and Cedar Paul (New York: Free Press, [1911] 1968), 191.
8. A ten-volume edition of his collected works (cited here as AWG) appeared from 1997 to 2003. A two-volume biography and a number of other volumes on him have been published in the last two decades. See Reinhard Blomert, *Intellektuelle im Aufbruch: Karl Mannheim, Alfred Weber, Norbert Elias und die Heidelberger Sozialwissenschaften der Zwischenkriegszeit* (Munich: Carl Hanser, 1999); Reinhard Blomert, Hans Ulrich Eßlinger, Norbert Giovannini (eds.) *Heidelberger Sozial- und Staatswissenschaften; Das Institut für Sozial- und Staatswissenschaften zwischen 1918 und 1958* (Marburg, Germany: Metropolis, 1997); Eberhard Demm, "Alfred Weber und sein Bruder Max," *Kölner Zeitschrift für Soziologie und Sozialpsychologie*, 35 (1983), 1–28; Eberhard Demm (ed.), *Alfred Weber als Politiker und Gelehrter* (Wiesbaden: Franz Steiner, 1986); Eberhard Demm, "Max and Alfred Weber in the Verein für Sozialpolitik," trans. Caroline Dyer, in Wolfgang J. Mommsen and Jürgen Osterhammel (eds.), *Max Weber and His Contemporaries* (London: Unwin Hyman, 1987), 8–105; Eberhard Demm, *Ein Liberaler in Kaiserreich und Republik. Der politische Weg Alfred Webers bis 1920* (Boppard, Germany: Harald Boldt, 1990); Eberhard Demm, *Von der Weimarer Republik zur Bundesrepublik. Der politische Weg Alfred*

Webers 1920–1958 (Düsseldorf: Droste, 1999); Eberhard Demm, *Geist und Politik im 20. Jahrhundert. Gesammelte Aufsätze zu Alfred Weber* (Frankfurt/Main: Peter Lang, 2000); Eberhard Demm (ed.), *Alfred Weber zum Gedächtnis: Selbstzeugnisse und Erinnerungen von Zeitgenossen* (Frankfurt/Main: Peter Lang, 2000); Volker Kruse, *Soziologie und "Gegenwartskrise." Die Zeitdiagnosen Franz Oppenheimers und Alfred Webers* (Wiesbaden, Germany: DUV, 1990); Hans G. Nutzinger (ed.), *Zwischen Nationalökonomie und Universalgeschichte. Alfred Webers Entwurf einer umfassenden Sozialwissenschaft in heutiger Sicht* (Marburg, Germany: Metropolis, 1995).
9. Two articles are representative of those in the English language that cite Alfred Weber at all. Lawrence Scaff briefly acknowledges Weber's contribution but is more interested in the development of his work by his colleague Emil Lederer. See Scaff, "Modernity and the tasks of a sociology of culture," *History of the Human Sciences*, 3 (1990), 85–100. Josef Bleicher not only ignores Weber in favor of his "student," Norbert Elias, but also goes on to say that Weber is to be avoided because of his association with the old mandarin ideal. See Bleicher, "Struggling with *Kultur*," *Theory, Culture & Society*, 7 (1990), 97–106. French sociologist Raymond Aron, whose short treatment of Alfred Weber was one of the very few available in English, drew a similar conclusion writing that Weber's cultural sociology "is the product of a revolt against civilization." See Aron, *German Sociology*, trans. Mary and Thomas Bottomore (New York: Free Press, 1964). More representative is Ralph Schroeder's study of Max Weber and the sociology of culture, which mentions Alfred only once, in a note. See Schroeder, *Max Weber and the Sociology of Culture* (London: Sage, 1992).
10. Jansen, although admiring of Demm's massive biography of Weber, laments that Demm wishes to view Weber "too much as an outstanding individual and too little as a typical representative of liberal human scientists, his generation, German political academics, etc. of that period." Jansen, "Neues von deutschen Gelehrtenpolitik," 29. The latter is what I do.
11. For a discussion of my earlier methodology, see Colin Loader, *The Intellectual Development of Karl Mannheim* (Cambridge, UK: Cambridge University Press, 1985), 1–9, 178–89.
12. Peter Wagner, *Sozialwissenschaften und Staat. Frankreich, Italien, Deutschland 1870–1980* (Frankfurt/Main: Campus, 1990). Erhart Stölting has devised a similar concept, "conceptual configuration," but I have chosen Wagner's because it emphasizes the reciprocity of science and other institutions. See Stölting, *Akademische Soziologie in der Weimarer Republik* (Berlin: Duncker & Humblot, 1986), 33–37.
13. Pierre Bourdieu, "The Force of Law: Toward a Sociology of the Juridical Field," trans. Richard Terdiman, *The Hastings Law Journal*, 38 (1987): 814–53; Pierre Bourdieu, *Outline of a Theory of Practice*, trans. Richard Nice (Cambridge, UK: Cambridge University Press, 1995); Pierre Bourdieu, *Homo Academicus*, trans. Peter Collier (Stanford: Stanford University Press, 1988); Pierre Bourdieu, *The Field of Cultural Production*, ed. Randal Johnson (New York: Columbia University Press, 1993); Pierre Bourdieu, *The Logic of Practice*, trans. Richard Nice (Stanford: Stanford University Press, 1990).
14. Bourdieu, *The Logic of Practice*, 5.
15. Bourdieu, *Outline*, 164–71, 177; "Force of Law," 823–25; *Homo Academicus*, 10–11.
16. Bourdieu, *Outline*, 191; *Homo Academicus*, xxv, 210.

17. Niklas Luhmann, *Social Systems* (Stanford: Stanford University Press, 1996), 460–61.
18. Ibid., 461–64. Bourdieu describes a similar scenario in *Homo Academicus*, 36.
19. Luhmann, *Social Systems*, 464.
20. For a similar discussion, see Seymour Chatman's structuralist narrative theory in *Story and Discourse* (Ithaca, NY: Cornell University Press, 1978), 19.
21. "Nationalökonomik" is usually translated as "political economy." I have chosen the more literal translation because of the School's emphasis on "politics" as a divisive sphere.
22. Georg Lukács, "Zum Wesen und zur Methode der Kultursoziologie," *Archiv für Sozialwissenschaft und Sozialpolitik*, 39 (1915), 217.
23. Alfred Weber, "Der Beamte," (1910), AWG 8: 98–117.
24. Demm, *Weimarer Republik zur Bundesrepublik*, 278.
25. Weber, *Kulturgeschichte als Kultursoziologie* (1935, 1950), AWG 1: 61.
26. For a harsher judgment, see Hans-Ulrich Wehler's review of the first volume of the AWG, *Kulturgeschichte als Kultursoziologie*. Wehler, "Reiter- und Immerweitervölker. Alfred Weber hat den Aufgalopp zur modernen Kulturgeschichte verpaßt," *Frankfurter Allgemeine Zeitung* 238 (Oct. 14, 1997), L36. Volker Kruse, in an exercise in the sociology of knowledge, argues that the "failure" of Weber's cultural sociology in the 1950s and after was due as much to its "receiver" (the prevailing empirical, quantitative sociology of the period) as to its "sender." Kruse, "Warum scheiterte Alfred Webers Kultursoziologie? Ein Interpretationsversuch," in Demm and Chamba (eds.), *Soziologie, Politik und Kultur* (Frankfurt/Main: Peter Lang, 2003), 219–29.
27. Wolf Lepennies, *The Seduction of Culture in German History* (Princeton: Princeton University Press, 2006), 9. For an earlier version of this approach, see Fritz Stern, "The Political Consequences of the Unpolitical German," in *History 3* (New York: Meridian, 1960), 104–34.

1 The Context of Alfred Weber's Early Work

1. Marianne Weber, *Max Weber: A Biography*, trans. Harry Zohn (New York: Wiley, 1975); Arthur Mitzman, *The Iron Cage: An Historical Interpretation of Max Weber* (New York: Grosset & Dunlap, 1969); Lawrence A. Scaff, *Fleeing the Iron Cage: Culture, Politics and Modernity in the Thought of Max Weber* (Berkeley & Los Angeles: University of California Press, 1989). For Alfred's own personal experience, see Eberhard Demm, *Ein Liberaler in Kaiserreich und Republik. Der politische Weg Alfred Webers bis 1920 (Boppard, Germany: Harald Boldt, 1990)*, 5–27.
2. Alfred Weber, "Professor Alfred Weber, Heidelberg," in Eberhard Demm, (ed.), *Alfred Weber zum Gedächtnis: Selbstzeugnisse und Erinnerungen von Zeitgenossen* (Frankfurt/Main: Peter Lang, 2000), 21–23.
3. Kevin Repp, *Reformers, Critics, and the Paths of German Modernity: Anti-politics and the Search for Alternatives, 1890–1914* (Cambridge, MA: Harvard University Press, 2000), 19–29, 227.
4. Ibid., 227.
5. Dennis Sweeney, "Governing the Social in Wilhelmine Germany: Rethinking German Modernity," American Historical Association Meetings, San Francisco, Jan. 6, 2000.

6. Peter Wagner, *Sozialwissenschaften und Staat. Frankreich, Itlaien, Deutschland 1870–1980* (Frankfurt/Main: Campus, 1990), 23–57.
7. Fritz K. Ringer, *The Decline of the German Mandarins* (Cambridge, MA: Harvard University Press, 1969).
8. This discussion follows the typology of Volker Kruse, *Soziologie und "Gegenwartskrise." Die Zeitdiagnosen Franz Oppenheimers und Alfred Webers* (Wiesbaden, Germany: Deutscher Universitäts Verlag, 1990), 31–34. Kruse actually used the term "sociologists," because his work points ahead to the Weimar Republic, when sociology made inroads as an established discipline at German universities.
9. Raymond Geuss, "Kultur, Bildung, Geist," *History and Theory* 35 (1996): 153.
10. Jürgen Habermas, *The Structural Transformation of the Public Sphere: An Inquiry into a Category of Bourgeois Society*, trans. Thomas Burger and Frederick Lawrence (Cambridge, MA: MIT Press, 1989). I stress that these are ideal types. Many, if not most, thinkers during this time wanted some combination of the two institutional forms. But while both might be advocated, one was privileged over the other.
11. Ibid., 23–28, 55–56, 62–64, 111.
12. Ibid., 72, 116–22.
13. Reinhard Kosselleck, "Einleitung: Zur anthropologischen und semantischen Struktur der Bildung," in Reinhard Kosselleck (ed.), *Bildungsgüter und Bildungswissen* (Stuttgart: Klett-Cotta, 1990); Klaus Vondung, "Zur Lage der Gebildeten in der wilhelminische Zeit," in Vondung (ed.), *Das wilhelminischen Bildungsbürgertum: Zur Sozialgeschichte seiner Ideen* (Göttingen, Germany: Vandenhoeck & Ruprecht, 1976); Rüdiger vom Bruch, *Wissenschaft, Politik und öffentliche Meinung: Gelehrtenpolitik im wilhelminischen Deutschland 1890–1914* (Husum, Germany: Matthiesen, 1980).
14. Louis Miller, "Between Kulturnation and Nationalstaat: The German Liberal Professoriate, 1848–1870," *German Studies Review*, Special Issue (1992): 33–54.
15. Wagner, *Sozialwissenschaften*, 132.
16. Christian Simon, *Staat und Geschichtswissenschaft in Deutschland und Frankreich 1871–1914* (Bern, Switzerland: Peter Lang, 1988), I, 6; Manfred Riedel, "Der Staatsbegriff der deutschen Geschichtsschreibung des 19. Jahrhunderts in seinem Verhältnis zur klassisch-politischen Philosophie," *Der Staat* 2 (1963): 49.
17. Charles McClelland, *The German Historians and England: A Study in Nineteenth Century Views* (Cambridge, UK: Cambridge University Press, 1971), 6; Rüdiger vom Bruch, "Historiker und Nationalökonomen in wilhelminischen Deutschland," in Klaus Schwabe (ed.), *Deutsche Hochschullehrer als Elite 1815–1945* (Boppard, Germany: Harald Boldt, 1988), 150; Dieter Lindenlaub, *Richtungskämpfe im Verein für Sozialpolitik: Wissenschaft und Sozialpolitik vom Beginn des "neuen Kurses" biz zum Ausbruch des ersten Weltkrieges 1890–1914* (Wiesbaden: Franz Steiner, 1967), 15.
18. The first generation, whose leading figures included Wilhelm Roscher, Karl Knies and Johannes Hildebrand, will not be discussed here.
19. Harald Winkel, *Die deutsche Nationalökonomie im 19.Jahrhundert* (Darmstadt, Germany: Wissenschaftliche Buchgesellschaft, 1977), 84–88.
20. Gustav Schmoller, *Über einige Grundfragen der Socialpolitik und der Volkswirtschaftslehre* (Leipzig: Duncker & Humblot, 1898), 324–25.

21. Samuel Bostaph, "The Methodological Debate between Carl Menger and the German Historicists," in Mark Blaug (ed.), *Carl Menger (1840–1921)* (Aldershot and Brookfield, VT: Edward Elgar, 1992), 137–50.
22. Schmoller, *Einige Grundfragen*, 336.
23. Schmoller, "Rede zur Eröffnung der Besprechung über die sociale Frage in Eisenach den 6. Oktober 1872," in Gustav Schmoller, *Zur Social- und Gewerbepolitik der Gegenwart: Rede und Aufsätze* (Leipzig: Duncker & Humblot, 1890), 9; Albert Müssiggang, *Die soziale Frage in der historischen Schule fer deutschen Nationalökonomie* (Tübingen: Mohr Siebeck, 1968), 225–27. The water metaphor was important to Historical School during this period as an expression of their opposition to the rigidity of classical national economy and Marxism. Even as he moved away from Schmoller's position, Alfred Weber would continue to use this metaphor with terms such as stream, current, and waves.
24. Schmoller, *Einige Grundfragen*, 322, 325.
25. Schmoller, *Einige Grundfragen*, 218. Pierangelo Schiera notes that England, France, and Italy did not have a corresponding concept to "German science." In proclaiming a national science, German economists demarcated their "system" from the "environment" of foreign science. The economic field without the spiritual guidance was derogated as "Manchesterism." Schiera, *Laboratorium der bürgerlichen Welt: Deutsche Wissenschaft im 19. Jahrhundert*, trans. Klaus-Peter Tieck (Frankfurt/Main: Suhrkamp, 1992), 10.
26. Gustav Schmoller, "The Origin and Nature of German Institutions," in Otto Hintze et. al., *Modern Germany in Relation to the Great War*, trans. William Wallace Whitelock (New York, 1916), 193, 213; Manfred Schön, "Gustav Schmoller and Max Weber," trans. Elizabeth King, in Wolfgang J. Mommsen and Jürgen Osterhammel (eds.), *Max Weber and His Contemporaries* (London: Unwin Hyman, 1987), 65; Lindenlaub, *Richtungskämpfe*, 24; Irmela Gorges, *Sozialforschung in Deutschland 1872–1914: Gesellschaftliche Einflüsse auf Themen- und Methodenwahl des Vereins für Sozialpolitik* (Königstein/Ts: Anton Hain, 1980), 330; Müssiggang, *Soziale Frage*, 156, 209.
27. Schmoller, "Rede zur Eröffnung," 9.
28. Gustav Schmoller, "Ueber Zweck und Ziele des Jahrbuchs," *[Schmollers] Jahrbuch für Gesetzgebung, Verwaltung und Volkswirtschaft im Deutschen Reich*, 1 (1881): 8–9; Schmoller, *Einige Grundfragen*, 340–42.
29. This term refers to a collective morality, as distinguished from *Moralität*, which is individual. "*Sitte*" is usually translated as "custom."
30. Schmoller, *Einige Grundfragen*, 337–38; Müssiggang, *Soziale Frage*, 95–98; Riedel, "Staatsbegriff," 43. Bruno Hildebrand and Johannes Conrad reported that at the first meeting of the Social Policy Association, Rudolf von Gneist was alone in claiming that economics and morality were opposites that could not be combined. And they qualified his position by noting that Gneist was a jurist, not a national economist. Bruno Hildebrand and Johannes Conrad, "Die Eisenacher Versammlung zur Besprechung der soziale Frage und Schmollers Eröffnungsrede," *Jahrbücher für Nationalökonomie und Statistik*, 20 (1873): 1.
31. Schmoller, *Einige Grundfragen*, 125–63; Gorges, *Sozialforschung*, 7, 57; Lindenlaub, *Richtungskämpfe*, 2.
32. Schmoller, "Rede zur Eröffnung," 11.
33. By the end of the empire, the role of the academics in this relationship had become less synthetic and more one-sided. Although Schmoller believed

that influence flowed mostly one way, from them to the state, their role had changed during the second half of the nineteenth century from that of participants in the formulation of government policy to that of merely publicists for governmental policy. They had become simply a vehicle by which the state addressed its public. Bruch, *Wissenschaft, Politik*, 137, 210.
34. Eckart Pankoke, *Sociale Bewegung—Sociale Frage—Sociale Politik: Grundfragen der deutschen "Sozialwissenschaft" im 19. Jahrhundert* (Stuttgart: Ernst Klett, 1970), 167; Wagner, *Sozialwissenschaften*, 87; Müssiggang, *Soziale Frage*, 11.
35. David F. Lindenfeld, *The Practical Imagination: German Sciences of State in the Nineteenth Century* (Chicago: University of Chicago Press, 1997), 295.
36. This is not to say that the academics of the Historical School were in complete agreement with one another or with specific policies of the state. Within the "universe of discourse" established by the coalition, Adolf Wagner challenged Schmoller from the right, while Lujo Brentano challenged him from the left. See Kenneth D. Barkin, *The Controversy over German Industrialization, 1890–1902* (Chicago: University of Chicago Press, 1970); James J. Sheehan, *The Career of Lujo Brentano: A Study of Liberalism and Social Reform in Imperial Germany* (Chicago: University of Chicago Press, 1966). Although there were differences in goals and methods, the centrality of the state-civil society dualism was not challenged, Wagner wanted an even stronger role for the state, while Brentano offered no viable political alternative to the bureaucratic state.
37. Schmoller, *Einige Grundfragen*, 343.
38. Kruse, "Von der historischen Nationalökonomie zur historischen Soziologie; Ein Paradigmenwechsel in den deutschen Sozialwissenschaften um 1900," *Zeitschrift für Soziologie* 19 (1990), 149–65.
39. Schmoller, *Einige Grundfragen*, 341.
40. Gunter Scholtz, *Zwischen Wissenschaftsanspruch und Orientierungsbedürfnis: Zu Grundlage und Wandel der Geisteswissenschaften* (Frankfurt/Main: Suhrkamp, 1991), 10, 31, 38–39, 47–48, 51: Erhart Stölting, *Akademische Soziologie in der Weimarer Republik* (Berlin: Duncker & Humblot, 1986), 25–26, 33.
41. Bruch, "Historiker": 105–6.
42. Barkin, *Controversy*.
43. Max Weber, "Developmental Tendencies in the Situation of East Elbian Rural Labourers," in Keith Tribe (ed.), *Reading Weber* (London: Routledge, 1989), 158–87.
44. David Blackbourn, "The Politics of Demagogy," *Past and Present* 113 (1986): 152–84.
45. Helmuth Schuster, *Industrie und Sozialwissenschaften: Eine Praxisgeschichte der Arbeits- und Industrieforschung in Deutschland* (Opladen, Germany: Westdeutscher Verlag, 1987), 61–99; Habermas, *Structural Transformation*, 151; Dieter Krüger, "Max Weber and the 'Younger' Generation in the Verein für Sozialpolitik, in Mommsen and Osterhammel, *Max Weber*, 71–72.
46. Schuster, *Industrie*, 82–83.
47. Schuster, *Industrie*, 84–85; Geuss, "Kultur": 157; Bruch et. al. (eds.), *Kultur und Kulturwissenschaften um 1900. Krise der Moderne und Glaube an die Wissenschaft* (Stuttgart: Franz Steiner, 1989), 11–12; Klaus Lichtblau, *Kulturkrise und Soziologie um die Jahrhundertwende: Zur Genealogie der Kultursoziologie in Deutschland* (Frankfurt/Main: Suhrkamp, 1996), 17–19; Woodruff D. Smith, *Politics and the Sciences of Culture in Germany, 1840–1920* (Oxford: Oxford University Press, 1991), 201.

48. Fritz Stern, *The Politics of Cultural Despair* (Garden City: Anchor Books, NY, 1965); Stephen E. Aschheim, *The Nietzsche Legacy in Germany, 1890–1990* (Berkeley & Los Angeles: University of California Press, 1992); Hans-Joachim Lieber, *Kulturkritik und Lebensphilosophie: Studien zur deutschen Philosophie der Hundertwende* (Darmstadt: Wissenschaftliche Buchgesellschaft, 1974); Lichtblau, *Kulturkrise*, 35. Kevin Repp notes the importance of various biological theories, especially vitalism, for this generation. These theories were often tied to issues of social reform. Kevin Repp, *Reformers*, 39–48.
49. Lindenlaub, *Richtungskämpfe*, 272–91.
50. Here I find myself in direct disagreement with Dirk Käsler, who also sees the themes of early German sociology as essentially dualistic with no discrepancy between the dualisms. Among the dualisms he lists as the dominant themes of the discipline are *Gemeinschaft-Gesellschaft*, irrationality-rationality, culture-civilization and state-society. See Käsler, *Die frühe deutsche Soziologie und ihre Entstheungs-Milieus 1909 bis 1934* (Opladen, Germany: Westdeutscher Verlag, 1984), 311.
51. Weber, "Der Beamte" (1910), AWG 8: 98–117.
52. An important strategy was to characterize their epoch as "capitalistic," a designation that most of Schmoller's generation rejected. Lindenlaub, *Richtungskämpfe*, 291–365. Schmoller's discomfort with this newer generation is evident in his "euphemistic" defensive alternative to the term "capitalism": "That which Werner Sombart called capitalism I would much rather designate as the modern monetary form of economic organization which developed under the liberal system of unrestricted occupational mobility, free competition and the unlimited desire for gain." Sombart, "Der kapitalistische Unternehmer." *Archiv für Sozialwissenschaft und Sozialpolitik* 29 (1909): 696.
53. Lindenfeld, *Practical Imagination*, 319; Repp, *Reformers*, 53–63.
54. Stern, *Politics of Cultural Despair*.
55. Georg Lukács called Alfred Weber "the most outstanding representative of this transitional form." Georg Lukács, *The Destruction of Reason*, trans. Peter Palmer (London: Merlin, 1980), 620.
56. The exception was Tönnies's *Gemeinschaft und Gesellschaft* (1887). However, Tönnies, a decade older than the other members of his generation, was a much more marginal figure in the university system than the Webers. And his most important works during the 1890s, with the exception of his study of Hobbes, were empirical studies of the Hamburg dock strikes, which Schuster describes as situational studies.

2 Early Economic Writings, 1897–1908

1. Eberhard Demm, *Ein Liberaler in Kaiserreich und Republik. Der politische Weg Alfred Webers 1920–1958* (Boppard, Germany: Harald Boldt, 1990), 28–29. Demm notes that Weber's positions on many matters were much closer to Lujo Brentano's than to Schmoller's. Weber later wrote to Brentano: "Among the whole great generation that went before me, I have actually only ever felt close to you." AWG 10: 110. Nevertheless, it was Schmoller who directed his early work.
2. Weber referred to these strikes in his dissertation, noting that Germany had joined the international wave against the sweating system. Weber, "Hau-

sindustrielle Gesetzgebung und Sweating-System in der Konfektionsindustrie" (1897), AWG 5: 27.
3. Demm, *Liberaler*, 33-34.
4. Weber, "Die volkswirtschaftliche Aufgabe der Hausindustrie; Akademische Antrittsvorlesung" (1901), AWG 5: 246.
5. Weber is given credit for being the first to see this relationship. See Karin Hausen, "Technical Progress and Women's Labour in the Nineteenth Century: the Social History of the Sewing Machine," in *The Social History of Politics*, ed. Georg Iggers (Leamington Spa, UK: Berg, 1985), 273.
6. Weber, "Die Hausindustrie und ihre gesetzliche Regelung: Referat" (1900), AWG 5: 225-27, 239.
7. Ibid., 225-27.
8. Weber, "Volkswirtschaftliche Aufgabe," AWG 5: 255-56.
9. Weber, "Hausindustrielle Gesetzgebung," AWG 5: 29-33; "Neuere Schriften über die Konfektionsindustrie" (1897), AWG 5: 95; "Die Entwickelung der deutschen Arbeiterschutzgesetzgebung seit 1890" (1897), AWG 5: 97; "Das Sweatingsystem in der Konfektion und die Verschläge der Kommission für Arbeiterstatistik" (1897), AWG 5: 63-65.
10. Weber, "Hausindustrielle Gesetzgebung," AWG 5: 55-58; "Neuere Schriften," AWG 5: 92; "Einleitung. Die Entwicklungsgrundlagen der großstätischen Frauenhausindustrie" (1899), AWG 5: 187, 195, 201-2; "Hausindustrie," AWG 5: 237-40; "Volkswirtschliche Aufgabe," AWG 5: 254.
11. For example, small farmers might continue to do some domestic labor during the winter.
12. Weber, "Hausindustrielle Gesetzgebung," AWG 5: 42-43; "Sweatingsystem," AWG 5: 76-78; "Entwicklungsgrundlagen," AWG 5: 204; "Der Arbeiterschutz in der Konfektion und verwandten Gewerben" (1899), AWG 5: 152; "Beschränkung der Heimarbeit in der Konfektionsindustrie" (1899), AWG 5: 154-55; "Hausindustrie," AWG 5: 243-45.
13. Weber, "Entwickelung," AWG 5: 98-99.
14. Ibid., 100, 112-13, 140.
15. Weber, "Hausindustrie," AWG 5: 238-39.
16. Weber, "Die Arbeitslosigkeit und die Krisen" (1902), AWG 5: 319-23, 329-33.
17. Weber, "Deutschland am Scheidewege" (1902), AWG 5: 335-45; "Deutschland am Scheidewege; Eine Replik" (1903), AWG 5: 346-59.
18. Weber, "Deutschland und der wirtschaftliche Imperialismus" (1904), AWG 5: 382-84, 387, 395-96.
19. Demm, *Liberaler*, 43
20. Weber, "Wirtschaftliche Imperialismus," AWG 5: 382-83.
21. Demm, *Liberaler*, 44-55; Else Jaffé, "Biographische Daten Alfred Webers (1868-1919)," and Astrid Lange-Kirchheim, "Alfred Weber und Franz Kafka," both in Demm, (ed.), *Alfred Weber als Politiker*, 189-91, and 113-49.
22. The three most important studies are typical of the literature as a whole. Roland Eckert completely ignores the book. Volker Kruse mentions it in passing but does not attempt to integrate it with Weber's other writings. Eberhard Demm provides a brief description and does connect it with Weber's political concerns, especially his support of a central European customs union, but does not tie it to the cultural sociology. See Eckert, *Die Geschichtstheorie Alfred Webers* (Tübingen, Germany: Kyklos, 1970); Volker Kruse, *Soziologie und "Gegenwartskrise." Die Zeitdiagnosen Franz Oppenheimers*

und Alfred Webers (Wiesbaden, Germany: Deutscher Universitäts-Verlag, 1990); Demm, *Liberaler.*
23. It was the first of his works to be translated and stayed in print much longer than the other translations.
24. Quoted in Leonore Gräfin Lichnowsky, "Alfred Weber's Standortslehre in ihrer politische Bedeutung" in *Alfred Weber als Politiker*, ed. Demm, 11–21.
25. Alfred Weber, *Theory of the Location of Industries* trans. Carl J. Friedrich (Chicago: University of Chicago Press, 1929), 2–3*, 28 [Weber, *Über den Standort der Industrien: 1, Reine Theorie des Standorts* (1909), AWG 6: 37–38, 59–60]. An asterisk (*) indicates that I have modified the translation.
26. Ibid., 22 [53].
27. Elizabeth Niederhauser, *Die Standortstheorie Alfred Webers; Studien über die Frage ihrer Gültigkeit und Fruchtbarkeit* (Weinfelden, Switzerland: Neuenschwander'sche Verlagsbuchhandlung, 1944), 189. When the book was reviewed in *Schmollers Jahrbuch*, it was assigned fairly to the Austrian economist Joseph Schumpeter and received a favorable review. See Schumpeter, Review of *Alfred Weber, Über den Standort der Industrien, [Schmollers] Jahrbuch*, 34 (1910), 1356–59.
28. Weber, *Theory of Location*, 5–6 [39–40]; Bernard W. Dempsey, *The Frontier Wage; The Economic Organization of Free Agents* (Chicago: Loyola University Press, 1960), 39.
29. Friedrich von Wieser, *Social Economics*, trans. A. Ford Hinrichs (New York: Augustus Kelley, 1967), 5–6.
30. Othmar Spann, *The History of Economics*, trans. Eden and Cedar Paul (New York: Norton, 1930), 175–76. Carl Menger, the Austrian who precipitated the "controversy over methods" with Schmoller in the 1880s, made similar statements. See Menger, *Investigations into the Method of the Social Sciences*, trans. Francis J. Nock (New York, SUNY Press, 1985), 78–79.
31. Othmar Spann described the difference thusly: "Can the laws of the internal structure and the evolution of political economy be studied as if they existed 'by themselves,' as if they formed a closed and self-determining system, originating out of purely economic causation (individual self-interest); or should we, rather, regard economics as inseparably interconnected with other provinces of society, and therefore not subject to laws peculiar to itself, but participating in the historically conditioned structure and development of society as a whole." Spann, *History*, 244.
32. Some friendly analysts have tried to play down that connection. Werner Sombart claimed that Weber's approach was more that of the Classical School than of the Austrian School in that he abstracted from a specific historical, capitalism, rather than postulating universal factors and laws. Sombart, "Einige Anmerkungen zur Lehre vom Standort der Industrieen," *Archiv für Sozialwissenschaft und Sozialpolitik*, 30 (1910), 750–51. That Weber did not agree with Sombart's assessment can be seen in his later article, where his summary of the first volume was titled "general theory," and more specific material, partly inspired by Sombart's review, was labeled "capitalistic theory." Weber, "Industrielle Standortslehre (Allgemeine und kapitalistische Theorie des Standortes)" (1923) AWG 6: 287–328. Niederhauser (*Standortstheorie*, 104–9) argues that Weber's method, like that of his brother Max, constructed ideal types. This would mean that his abstractions were heuristic and ultimately historical, not universal. Again, this would seem to contradict Weber's own

statements that he was "restricting" himself to a "general, a priori theory." Weber, "Industrielle Standortslehre," 290.
33. Weber, "Vorwort," in Otto Schlier, *Der deutsche Industriekörper seit 1860; Allgemeine Lagerung der Indistrie und Industriebezirksbildung* (1922), AWG 6: 329–33.
34. Weber, *Theory of Location*, 12, 17, 29–35 [45, 48, 60–69]; Walter Isard, "Distance Inputs and the Space Economy: The Local Equilibrium of the Firm," in *Spatial Economic Theory*, ed. Robert D. Dean, et al. (New York: Free Press, 1970), 34; August Lösch, *The Economics of Location*, trans. William H. Woglom and Wolfgang F. Stolper (New Haven: Yale University Press, 1954), 5.
35. For example, in the processing of an ore, the final product—the pure metal—weighed much less than the ore itself, i.e., the material index was greater than 1, which meant that transporting the processed metal from the smelter to the place of consumption would be less expensive than transporting the ore from the mine to the smelter. Therefore, the smelter would be located as close to the mine and away from the place of consumption as possible.
36. Weber, *Theory of Location*, 54, 75 [83, 101–2].
37. Ibid., 102–4 [126–28], quote on 103 [126].
38. Ibid., 137 [155].
39. Ibid., 125, 138 [145, 156].
40. Ibid., 173, 213 [184, 219]; Edwin von Böventer, "Towards a Unified Theory of Spatial Economic Structure," in *Spatial Economic Theory*, ed. Dean, et al., 326.
41. Weber, *Theory of Locations*, 213* [219].
42. Ibid., 21–22* [52].
43. Ironically, at one place in the above passage in the 1929 translation, "*Kultur*" is rendered as "civilization."
44. Ibid., 214–21 [220–26].
45. Ibid., 218* [223–24].
46. Ibid., 218–19* [224].
47. Sombart, "Einige Anmerkungen"; Weber, "Industrielle Standortslehre," AWG 6: 316–23.

3 Heidelberg and the Empire, 1907–1917

1. Wilhelm Heinrich Riehl, *Die Naturgeschichte des deutschen Volkes*, abridged edition (Leipzig: Reclam, n.d.), 90–103.
2. Alon Confino, *The Nation as a Local Metaphor: Württemberg, Imperial Germany, and National Memory, 1871–1918* (Chapel Hill: University of North Carolina Press, 1997).
3. Hubert Treiber and Karol Sauerland, (eds.), *Heidelberg im Schnittpunkt intellektueller Kreise* (Opladen: Westdeutscher Verlag, 1995), 9, 12.
4. Christian Jansen, *Professoren und Politik. Politisches Denken und Handeln der Heidelberger Hochschullehrer 1914–1935* (Göttingen: Vandenhoeck & Ruprecht, 1992), 14.
5. See Simmel, "The Metropolis and Mental Life," in Georg Simmel, *The Sociology of Georg Simmel*, ed. Kurt H. Wolff (New York: Free Press, 1950), 409–24. For the physiognomy of Berlin, see Peter Fritzsche, *Reading Berlin 1900* (Cambridge, MA: Harvard University Press, 1996). Also, Gerhard Masur, *Imperial Berlin* (New York: Dorset, 1989), 187–201.
6. Karl Mannheim, *Sociology as Political Education*, trans. and ed. David Kettler and Colin Loader (New Brunswick: Transaction, 2001), 89–90; M. Rainer

Lepsius, "Ach, Heidelberg: Beziehungsvolle Vergangenheiten," *Soziologische Revue*, 18 (1995): 477; Helene Tompert, *Lebensformen und Denkweisen der akademischen Welt Heidelbergs im wilhelminischen Zeitalter* (Lübeck & Hamburg: Matthiesen, 1969), 11, 17; Edmund W. Gilbert, *The University Town in England and West Germany* (Chicago: University of Chicago Press, 1961), 39–40, 47.
7. Jansen, *Professoren*, 32.
8. Quoted in Ibid.
9. Alfred Weber, "Die Jugend und das deutsche Schicksal: Persönliche Rückblicke und Ausblicke," in *Wegweiser in der Zeitwende*, ed. Elga Kern (Munich, Germany; Basel, Switzerland: Ernst, 1955), 62.
10. Salaries at Heidelberg could be twice those at Berlin. Tompert, *Lebensformen*, 36.
11. Robert E. Norton, *Secret Germany: Stefan George and His Circle* (Ithaca: Cornell University Press, 2002), 459.
12. Marianne Weber, *Max Weber: A Biography*, trans. Harry Zohn (New York: Wiley, 1975), 320, translation modified.
13. Tompert, *Lebensformen*, 25, 31.
14. Walter Z. Laqueur, *Young Germany: A History of the German Youth Movement* (New York: Basic, 1962), 5–6.
15. Lichtblau, *Kulturkrise*, 408. This criticism was reiterated by Karl Mannheim after the war. Mannheim, *Sociology*, 99–104.
16. Michael Stephen Steinberg, *Sabers and Brownshirts: The German Students' Path to National Socialism, 1918–1935* (Chicago: University of Chicago Press, 1977), 24; Norton, *Secret Germany*, 459.
17. In addition to the circles and groups in which Weber participated, and which are discussed below, there were many others. The "Heidelberg Community," organized by the art historian Wilhelm Franger, included the novelist Carl Zuckmayer, the socialist politician Carlo Mierendorff, the psychologist Hans Prinzhorn, but only one academic. Various artists and popular philosophers (such as Weber's friend Hermann Keyserling) were invited to give presentations. The Natural History-Medical Association dated from the mid-nineteenth century and included mainly academics, as did the Philosophical-Historical Association. The Baden-Badener Society was a philosophical-theological discussion circle. Incalcata (the Decalcified Academy) attracted young lecturers, including the historian Hajo Holborn and the philologist of India Heinrich Zimmer. Another group was organized by the popular novelist Emil Ludwig. Jansen, *Professoren*, 40–41.
18. Marianne Weber, "Academic Conviviality," *Minerva* 15 (1977): 220; Paul Honigsheim, "Der Max-Weber-Kreis in Heidelberg," *Kölner Vierteljahrshefte für Soziologie* 5 (1926): 271–72; Stefan Breuer, "Das Syndicat der Seelen. Stefan George und sein Kreis," and Rainer Kolk, "Das schöne Leben. Stefan George und sein Kreis in Heidelberg," both in *Heidelberg im Schnittpunkt*, ed. Treiber and Sauerland, 328–75 and 310–27; Alfred Weber, "Jugend," 62; Tompert, *Lebensformen*, 42.
19. Jansen, *Professoren*, 14, 33. The main party of the Heidelberg faculty was the National Liberal Party, which largely supported the policies of the imperial government. Tompert, *Lebensformen*, 61.
20. Quoted in Jansen, *Professoren*, 34.
21. Mannheim, *Sociology*, 84–85. For a similar argument concerning nationalism, see Confino, *Nation as a Local Metaphor*.
22. Wilhelm von Humboldt, "On the Historian's Task," *History and Theory* 6 (1967): 103–8; Ernst Troeltsch, *Der Historismus und seine Probleme*, vol. I: *Das*

logische Problem der Geschichtsphilosophie (Tübingen, Germany: Siebeck Mohr, 1922), 38, 252–53; Eduard Spranger, *Wilhelm von Humboldt und die Reform des Bildungswesens* (Tübingen, Germany: Max Niemeyer, 1960) 45–46, 65; Peter Hanns Reill, "Science and the Construction of the Cultural Sciences in Late Enlightenment Germany: The Case of Wilhelm von Humboldt," *History and Theory* 33 (1994): 350, 354, 363–64. For extended versions of this discussion, see Colin Loader, *Intellectual Development*, 17–19, and Loader and Kettler, *Karl Mannheim's Sociology as Political Education*, New Brunswick: Transaction, 2002), 19–46.
23. Loader and Kettler, *Karl Mannheim's Sociology*, 22–23.
24. Quoted in Jansen, *Professoren*, 33.
25. Michael Buselmeier, "Mythos Heidelberg," in *Auch eine Geschichte der Universität Heidelberg*, ed. Karin Buselmeier et al. (Mannheim: Edition Quadrat, 1985), 491–500.
26. Mannheim, *Sociology*, 85–86.
27. Kruse, *Soziologie*, 236. During this period, Weber himself agreed with this assessment, describing it as his "classical period" in which he was at the "zenith" of his productivity. In a much later list of his major works, however, none, with exception of *The Location of Industries* (which originated in the earlier period), from this period is mentioned. See Weber, "Der Mann von 40 Jahren" [1911] and "Lebenslauf September 1947," both in *Alfred Weber zum Gedächtnis*, ed. Demm, 51–52 and 21–23.
28. Max had resigned his professorship due to fragile health. George's only tie to the university came through his disciple Friedrich Gundolf, who held a university position. Norton, *Secret Germany*, 461–62, 476.
29. Alfred purposely moved into a house on the opposite side of the city from Max, who shared their parental house with Ernst Troeltsch. Mitzman, *Iron Cage*, 285–86; Marianne Weber, *Max Weber*, 449).
30. Martin Green, *The von Richthofen Sisters: The Triumphant and the Tragic Modes of Love* (New York: Basic, 1974), 129–30. Demm writes that Alfred was the victor almost from the beginning. Demm, "Eine Sexbombe der Jahrhundertwende. Else Jaffé-von Richthofen und ihre Liebhaber," in "*...da liegt der riesige Schatten Freud's jetzt nicht mehr auf meinem Weg.*" *Die Rebellion des Otto Gross*, ed. Raimund Dehmlow, et al. (Marburg, Germany: TransMIT, 2008). Also see Sam Whimster and Gottfried Heuer, "Otto Gross and Else Jaffé and Max Weber," *Theory, Culture and Society* 15 (1998), 129–60; Sam Whimster, "No place for a revolutionary. Heidelberg and Otto Gross," in *Sexual Revolutions: Psychoanalysis, History and the Father*, ed. Gottfried Heuer (Hove: Routledge, 2011), 181–200.
31. In 1909 Schmoller wrote to a colleague: "There is great indignation about the Webers among those on the right that I cannot share. They are neurotic; but they are the yeast and stimulus of our meetings; they are honorable people and extremely great talents. They certainly act as explosives for our *Verein*, that *meo voto* one must, however, hold together as long as possible." In Edgar Salin, *Lynkeus: Gestalten und Probleme aus Wirtschaft und Politik* (Tübingen, Germany: Mohr Siebeck, 1963), 62.
32. Demm has calculated that the pages containing Alfred's contributions to the debates slightly exceed in number those of Max. "Max and Alfred Weber," 96–97.
33. In 1907, the Weber brothers and Heinrich Herkner proposed a study of the impact of modern industry on workers. Alfred was more interested in the

impact of the work on the lives and personalities of the individual, while Max was more interested in worker productivity. The interviews with workers took place from 1908 until 1911 and resulted in six volumes. At the Association meeting in 1909, Alfred delivered an attack on the bureaucracy, a theme that would remain a life-long concern. Alfred was especially concerned with the impact of bureaucracy on the individual personality. Demm, "Max and Alfred Weber," 88–95.

34. Eberhard Demm, "Alfred Weber und sein Bruder Max," *Kölner Zeitschrift für Soziologie und Sozialpsychologie*, 35 (1983), 4; Eberhard Demm, *Ein Liberaler in Kaiserreich und Republik. Der politische Weg Alfred Webers bis 1920* (Boppard, Germany: Harald Boldt, 1990), 56; Lawrence A. Scaff, *Fleeing the Iron Cage: Culture, Politics, and Modernity in the Thought of Max Weber* (Berkeley and Los Angeles: University of California Press, 1989), 108–11

35. The merit of Talcott Parsons's rendering of Weber's *"stahlhartes Gehäuse"* as "iron cage" has been a matter of much debate. Stephen Kalberg in his retranslation of *The Protestant Ethic and the Spirit of Capitalism* (Los Angeles: Roxbury, 2002) uses a more literal "steel-hard casing." Peter Baehr offers "shell as hard as steel" in an essay that reviews the controversy. Baehr, "The 'Iron Cage' and the 'Shell as Hard as Steel': Parsons, Weber, and the *Stahlhartes Gehäuse* Metaphor in *The Protestant Ethic and the Spirit of Capitalism*," *History and Theory* 40 (2001), 153–69. While their versions are certainly more accurate depictions of what Weber meant, it is hard to see them capturing the academic imagination the way Parsons's has. One need only think of the many books that incorporate "iron cage" into their titles. Because Alfred Weber also used the term *"Gehäuse"* with similar connotations, I will translate it as "casing."

36. Colin Loader and Jeffrey Alexander, "Max Weber on Churches and Sects in North America: An Alternative Path toward Rationalization," *Sociological Theory*, 3 (1985): 1–6; David Beetham, *Max Weber and the Theory of Modern Politics* (Cambridge, UK: Polity, 1985); Harvey Goldman, *Max Weber and Thomas Mann: Calling and the Shaping of the Self* (Berkeley & Los Angeles, University of California Press, 1988).

37. Gerhard Lauer, *Die verspätete Revolution, Erich von Kahler. Wissenschaftsgeschichte zwischen konservativer Revolution und Exil* (Berlin: Walter de Gruyter, 1995), 183–98, 210–12, 239; Norton, *Secret Germany*, 457, 465, 484. When his disciple Gundolf published his book on Shakespeare after becoming a professor, George announced that "the professors have ruined him." Norton, *Secret Germany*, 465.

38. Marianne Weber, *Max Weber*, 455–64; Mannheim, *Sociology*, 90–97; Edgar Salin, *Um Stefan George* (Godesberg, Germany: Helmut Küpper, 1948), 157–62; Scaff, *Fleeing*, 106–8; Mitzman, *Iron Cage*, 261–71; Norton, *Secret Germany*, 478; Wolf Lepenies, *Sociology Between Literature and Science*, trans. R.J. Hollingdale (Cambridge, UK: Cambridge University Press, 1988), 198; Lauer, *Verspätete Revolution*, 247–49. Kahler's attack, which appeared in the year of Max Weber's death, produced a series of responses that have been collected and partially translated in *Max Weber's Science as a Vocation*, ed. Peter Lassman et al. (London: Unwin Hyman, 1989), 35–46. Alfred Weber did not participate in this dispute.

39. Demm, *Liberaler*, 58; Demm, *Geist und Politik*, 99–109.

40. Christa Dericum, "Einleitung," in Alfred Weber, *Haben wir Deutschen nach 1945 versagt? Politische Schriften: Ein Lesebuch*, ed. Dericum (Munich and Zurich: Piper, 1979), 18.

41. Demm, *Liberaler*, 55–56; Demm, *Weimarer Republik*, 80; Blomert, *Intellektuelle*, 148; Lauer, *Verspätete Revolution*, 227–29. Alfred never exerted the forcefulness over his circle that Max Weber and George exerted over theirs. He was more of a moderator than a leader.
42. An example is the account of Edgar Salin, a disciple of George and Alfred Weber's choice to fill a chair in his institute in the mid-1920s, of an appearance by Max Weber one evening. Max gave a lengthy impromptu speech along the lines of his later "Science as a Vocation," in which he drastically restricted the role of science in deciding cultural values. When Salin asked him how this restriction of science would apply to Theodor Mommsen's Roman history, Weber replied, "that is not science." To this Salin responded: "then I don't know how your science is able to serve anything that is alive and why it should be of interest to us." Salin, *Um Stefan George*, 162–63.
43. Laqueur, *Young Germany*, 32–38.
44. It is unclear what direct role, if any, Weber had is this organization. Joan Campbell lists him as an avid supporter. Campbell, *The German Werkbund* (Princeton: Princeton University Press, 1978), 105. And he delivered a speech to one of its conventions in 1928. But there is no mention of the group in Demm. Nevertheless, Weber was closely connected to Friedrich Naumann, who was a founder of the *Werkbund*, and had to have some ties, if only indirect, to it during World War I, when both he and it were supporters of Naumann's "Mitteleuropa" policy. At the end of the war, he mentioned the *Werkbund* favorably. "Die Bedeutung der geistigen Führer in Deutschland" (1918) AWG 7: 355.
45. Carl Zuckmayer claimed the Keyserling's *Travel Diary of a Philosopher* was almost as widely read as Oswald Spengler's *Decline of the West*. Zuckmayer, *Als wär's ein Stück von mir* (Vienna: Fischer, 1967). 298.
46. Hermann Keyserling, *The World in the Making*, trans. Maurice Samuel (New York: Harcourt Brace, 1927), 34–35; Demm, *Liberaler*, 55–64, 143–44; Beate Riesterer, "Alfred Weber's Position in German Intellectual History," in *Alfred Weber als Politiker*, ed. Demm, 91.
47. Peter Fritzsche, *German into Nazis* (Cambridge, MA: Harvard University Press, 1998).
48. This is especially true of a wartime collection of essays by many well-known German academics, including Schmoller. Otto Hintze et al., *Modern Germany in Relation to the Great War* (New York: Mitchell Kennerley, 1916).
49. Sombart, *Händler und Helden: Patriotische Besinnungen* (Munich: Duncker & Humblot, 1915).
50. Klaus Böhme, (ed.) *Aufrufe und Reden deutscher Professoren im Ersten Weltkrieg* (Stuttgart: Reclam, 1975), 47–49; Klaus Schwabe, *Wissenschaft und Kriegsmoral: Die deutschen Hochschullehrer und die politischen Grundfragen des Ersten Weltkrieges* (Göttingen: Musterschmidt, 1969), 21–22.
51. Böhme, *Aufrufe*, 3–43.
52. Schwabe, *Wissenschaft*, 21–29.
53. Jansen, *Professoren*, 109–11.
54. The only exception to this was the response to a critique by Georg Lukács, which is discussed in the next chapter.
55. Weber, *Gedanken zur deutschen Sendung* (1915), AWG 7: 116–77.
56. Demm has written extensively on Weber's wartime political activities. I will present only those elements than are relevant to his intellectual career in general. Demm, *Liberaler*, 152–255.

57. See Weber, *Gedanken*, 27, and "Die Bedeutung der geistigen Führer in Deutschland" (1918), AWG 7: 347–67.

4 The Question and Sociology of Culture

1. It has survived only as a set of student notes, and so I will cite it only sparingly. Weber, AWG 8: 263–314; Eberhard Demm, *Ein Liberaler in Kaiserreich und Republik. Der politische Weg Alfred Webers bis 1920* (Boppard, Germany: Harald Boldt, 1990), 138–39. Friedrich Gundolf wrote to Hermann Keyserling's wife Leonie: "Alfred Weber has given a critique of our era and its organization in his lecture course on cultural problems of the present, compared to which many of the discourses in our *Jahrbuch* are mild and tame; and the public listened devoutly and attentively" (in Demm, *Liberaler*, 139).
2. Weber, "Der Kulturtypus und seine Wandlung" (1909/10), AWG 8: 91.
3. In the quoted passage Weber equated "position" with "vocation" (*Beruf*). Within three years vocation would be treated as almost the opposite of position and more akin to cultivation. See especially Weber, "Das Berufsschichsal der Industriearbeiter" (1912), AWG 8: 344–68.
4. Weber, "Kulturtypus," AWG 8: 91–92.
5. Ibid., 93, 95–96; Weber, "Theodor Mommsen" (1906), AWG 8: 88. On the other hand, one could not go too far in the opposite direction and follow the lead of the "cultural aesthete," who sought to divorce himself from the everyday world. This type deprecated all political concerns and communal interests to escape into "aristocratic twaddle." If the bureaucrat wanted to turn workers into narrow people like himself, the cultural aesthete wanted nothing at all to do with them. On the head of this irresponsible type, Weber said, one should simply place a "fool's cap." Weber, "Kulturtypus," 94–95. Weber was clearly thinking here of the George Circle.
6. Quoted in Demm, *Liberaler*, 143.
7. Academic German references, such as Dilthey and Simmel, were missing in Weber's writings during this period, as was specific reference to Driesch. Nietzsche and Bergson were cited. See, for example, Weber, *Religion und Kultur* (1912), AWG 8: 318, 325–26.
8. Like Romanticism, vitalism had its conservative and radical poles. On the one hand, traditionalists used vitalist language to attack the spread of modernist movements, especially positivism. Radicals, on the other hand, used the same language to promote change and to attack rigid, undynamic form. "Depending upon the author, *Lebensphilosophie* could be used to support either of these positions, and in the case of more complex authors, both at the same time." Peter Hans Reill, "Science and the Construction of the Cultural Sciences in Late Enlightenment Germany: The Case of Wilhelm von Humboldt," *History and Theory* 33 (194), 345–66.
9. Sanford Schwartz, "Bergson and the Politics of Vitalism," in *The Crisis in Modernism: Bergson and the Vitalist Controversy*, ed. Frederick Burwick and Paul Douglass (Cambridge, UK: Cambridge University Press, 1992), 278–79.
10. Bergson, *Creative Evolution*, trans. Arthur Mitchell (New York: Random House, 1944), 32, 109, 146, 244, 290, 328, 340; Bergson, *An Introduction to Metaphysics*, trans. T. E. Hulme (Indianapolis: Harper, 1955), 41, 49–50.
11. Bergson, *Metaphysics*, 25–31.

12. Hans Driesch, *The Science and Philosophy of the Organism*, second edition (London: Black, 1929), 225, 262, 329; Hans Driesch, *The History and Theory of Vitalism*, trans. C. K. Ogden (London: Macmillan, 1914), 205.
13. Driesch, *History*, 205; Driesch, *Science*, 257, 264.
14. Horst H. Freyhofer, *The Vitalism of Hans Driesch: The Success and Decline of a Scientific Theory* (Frankfurt/Main: Peter Lang, 1982), 91–92.
15. Driesch, *History*, 199–201.
16. Driesch, *Science*, 78.
17. Bergson, *Creative Evolution*, 294; Driesch, *Science*, 244, 307.
18. Bergson, *Creative Evolution*, 194–96. Bergson wrote: "By intuition is meant the kind of *intellectual sympathy* by which one places oneself within an object in order to coincide with what is unique in it and consequently inexpressible. Analysis, on the contrary, is the operation which reduces the object to elements already known, that is, to elements common both to it and other objects. To analyze, therefore, is to express a thing as a function of something other than itself." Analytical knowledge, which includes both empiricism and rationalism, is always relative. It is confined to the crystallized fragments on the surface of reality. Only through intuition can one penetrate to the absolute unity of duration. *Metaphysics*, 21–24, 35, 37–38, 50–51.
19. Ibid., 45, 59; Bergson, *Creative Evolution*, xix–xx, 51, 153–54, 163–78.
20. Bergson, *Metaphysics*, 50–51, 42.
21. In choosing biology for his inspiration, Weber, whether intentionally or not, avoided the conventional human sciences–natural sciences dichotomy.
22. Weber, *Religion und Kultur (1912)*, AWG 8: 317.
23. This denial of life is true not only of religions like chiliastic Christianity and Buddhism, which focus on the escape from this world to another one, but also more "worldly" ones like Puritanism, which still subordinates this world to another and, in the words of his brother, has given rise to the iron cage of capitalism. Ibid., 319–22, 328–34.
24. Ibid., 334–35. While the civilization-culture dualism was not used explicitly in this lecture, Weber did use it elsewhere at this time, beginning with his lecture course. AWG, 8: 268–71.
25. Weber, *Religion und Kultur*, AWG 8: 335–36.
26. Ibid., 325–26.
27. Weber, Letter to Keyserling, AWG 10: 501.
28. Weber, *Religion und Kultur*, AWG 8: 337.
29. The same year in which he made his Prague speech, Weber discussed what was necessary for the "productive spirit," that is, the new cultural person. Such a person, he wrote, synthesized two others, the systematizer and the fantasist. The former of these two was grounded in reality, which he attempted to conceptualize in order to organize life. The latter, conversely, always strove for something new, wanting to experience the unknown and enjoy the creativity of life. Weber, "Der productive Geist" (1911), AWG 8: 254–59. Weber believed that both aspects, which were akin to the vitalists' dualism, were necessary for spiritual productivity.
30. Weber, "Letter to Keyserling," AWG 10: 501.
31. Weber, "Religion und Kultur," AWG 8: 338.
32. Weber, "Schule und Jugendkultur" (1913), AWG 7: 83–90.
33. Ibid.

34. Weber, "Rede über die freideutsche Jugendbewegung auf der Aufklärungversammlung in der Münchener Tonhalle am 9. Febr. 1914 nebst Schlusswort" (1914), AWG 7: 96.
35. Ibid., 95–101; Weber, "Eine Zuschrift" (1914), AWG 7: 106–7.
36. Weber, "Geleitwort" to Hans Staudinger, *Individuum und Gemeinschaft in der Kulturorganisation des Vereins* (1913), AWG 8: 373–78; Weber, "Individuum und Gemeinschaft" (1914), AWG 8: 381–83; Staudinger, "Das Kulturproblem und die Arbeiterpsyche," *Die Tat* 5 (1914): 990–1002.
37. Demm, *Liberaler*, 64.
38. Staudinger, *Individuum und Gemeinschaft in der Kulturorganisation des Vereins* (Jena: Eugen Diederichs, 1913), 9, 41, 48–49, 128–31.
39. Ibid., 137, 144; Staudinger, "Kulturproblem," 993–94, 1001–2.
40. Weber, "Geleitwort," AWG 8: 373–77; Weber, "Individuum," AWG 8: 381–83.
41. Weber, "Individuum," AWG 8: 383.
42. Ibid., 382–83; Weber, "Geleitwort," AWG 8: 375–76.
43. Weber, "Individuum," AWG 8: 381–82; Weber, "Geleitwort," AWG 8: 375.
44. Weber, "Individuum," AWG 8 : 382.
45. Weber, "Geleitwort," AWG 8 : 375–78.
46. Ibid., 377–78; Weber, "Individuum," AWG 8 : 383.
47. Weber, "Individuum," AWG 8 : 383.
48. Weber, "Program einer Sammlung 'Schriften zur Soziologie der Kultur'" (1913/14), AWG 8: 379–80; Weber, "Entgegnung" (1915), AWG 8: 384.
49. Roland Eckert, *Die Geschichtstheorie Alfred Webers* (Tübingen, Germany: Kyklos, 1970), 9.
50. Weber, AWG, 8: 268, 270.
51. Eberhard Demm has argued that the influence of the two vitalists was successional, with Driesch eventually replacing Bergson. Demm, *Liberaler*, 143–44. To a certain extent, this is correct in that one sees more of the Bergsonian approach in these early pieces. Weber did not use Driesch's "entelechy" or "constellation" in these works, although he did use the latter term in his article on Social Policy the same year. Weber, "Neuorientierung in der Sozialpolitik?" (1913), AWG 5: 475–85. And he used Driesch's term "*Anlage*" (self-changing potential) once in his address on cultural sociology. As I will argue below, the reason for Bergson's prewar prominence is that Weber was more concerned with establishing the dualism than he was with elaborating on the cultural side of it.
52. Weber, "Soziologische Kulturbegriff," AWG 8: 62.
53. Ibid., 62–64, 70–73.
54. Ibid., 68.
55. Ibid., 70.
56. Weber, "Entgegnung," AWG 8: 386.
57. Ibid.
58. Loader and Jeffrey Alexander, "Max Weber on Churches and Sects in North America: An Alternative Path toward Rationalization," *Sociological Theory* 3 (1985), 1–6.
59. Walter Köhler, "Die Ergebnisse des zweiten deutschen Soziologentages," *Schmollers Jahrbuch* 38 (1914), 415–36; Georg Lukács, "Zum Wesen und zur Methode der Kultursoziologie," *Archiv für Sozialwissenschaft und Sozialpolitik* 39 (1915), 216–22. In his response to Lukács, Weber stated that this speech had generated a great deal of opposition, but cited only Köhler. Weber, "Entgegnung," AWG 8: 384.

60. Lukács, "Zum Wesen," 218. In a later polemical work Lukács was harsher. Weber, he wrote, "attempted a 'synthesis', an intellectual 'illumination' of the irrational but without rationalizing it, a scientific approach that was antiscientific." Lukács, *The Destruction of Reason*, trans. Peter Palmer (London: Merlin, 1980), 621.
61. Köhler, "Ergebnisse," 417–18.
62. Lukács, "Zum Wesen," 218–19.
63. Weber, "Entgegnung," AWG 8: 386. Although Weber did not directly respond to Köhler, his rebuttal to Lukács applied to Köhler's critique as well.
64. That Köhler did not understand or accept Weber's refusal to conflate can be seen in two specific criticisms he made. He agreed with Weber that the work of art was subjective but questioned how the idea could be. And he also questioned how Weber could classify the state, law, and economy as forms of biological life rather than cultural phenomena. Köhler, "Ergebnisse," 422–23. But Weber had designated two different kinds of ideas, the first of which (*Idee*) was subjective and was assigned to the cultural sphere, the second, the thought (*Gedanke*), was objective and was assigned to the civilizational process. Thus, not all ideas were subjective; only those that were created from the person's inner being. Köhler simply did not recognize this dichotomy. Weber assigned the state, law, and economy to the civilizational sphere, the basis of which was satisfying biological needs, because he saw them as the means for the instrumental realization of material goals. This did not mean that he believed they had no cultural elements, but rather that they did not serve the basic cultural purpose of establishing meaning and values. This position was a continuation of his attack on the discursive coalition and its interpretation of the state.
65. Weber, AWG 8 : "Entgegnung," AWG 8: 386.
66. Weber, "Religion und Kultur," AWG 8: 337–38.
67. Weber, "Entgegnung," AWG 8: 387.
68. Weber, "Program einer Sammlung," AWG 8: 856. Among the proposed studies in these categories were: (1) Plastic arts in an industrial city, the social ideas of Richard Wagner, cinema as a capitalistic enterprise, modern theater and concert enterprise (Staudinger's book also fits into this category); (2) Worker intellectuals of a German industry, literary interests of a large German city, the social origins of leading talents; (3) The question of German officials, English bureaucracy, the Oxford Movement, the social importance of the Salvation Army. Weber, "Leben und Kultur. Beiträge zur Erkenntnis ihrer inneren Beziehung." Unpublished typescript. NL 197 Alfred Weber 14, Bundesarchiv Koblenz. This is a draft of "Program einer Sammlung."
69. For the tensions between assimilationists and Zionists see Jehuda Reinharz, *Fatherland or Promised Land: The Dilemma of the German Jew, 1893–1914* (Ann Arbor: University of Michigan Press, 1975); the most thorough and balanced account of Sombart's role is Friedrich Lenger, *Werner Sombart 1863–1941* (Munich: Beck, 1994), 187–218; also see Loader, "Werner Sombart's The Jews and Modern Capitalism," *Society* 39 (2001), 71–77.
70. Weber, untitled contribution in Werner Sombart et al, "Judentaufen" (1912), AWG 8: 339–41.
71. Ibid., 341–42.
72. Here Weber rejected the claim of Jewish assimilationists that they were full participants in the German culture whose religion happened to be Judaism

rather than Catholicism or Lutheranism. In other words, he denied that Judaism could be assimilated into German culture.
73. Ibid., 343.

5 The Cultural Theory of Politics

1. Schmoller, "Fürst Bülow und die preußisch-deutsche Politik im Frühjahr 1907," in Gustav von Schmoller, *Charakterbilder* (Munich & Leipzig: Duncker & Humblot, 1913), 101.
2. Ibid., 107–8, quote on 108. Also see Eberhard Demm, *Ein Liberaler in Kaiserreich und Republik. Der politische Weg Alfred Webers bis 1920* (Boppard, Germany: Harald Boldt, 1990), 85.
3. Weber, "Konstitutionelle oder parlamentarische Regierung in Deutschland?" (1907), AWG 7: 33.
4. Ibid., 34.
5. Ibid., 35–38.
6. Several years later (1912) Schmoller as much as admitted this, writing in the same newspaper, writing that while he still believed in the superiority of the bureaucratic state, "I am also conscious of the fact that those who so think are dominant in neither public opinion nor in political parties and least of all in the social classes." Schmoller, *Zwanzig Jahre Deutscher Politik, 1897–1917. Aufsätze und Vorträge* (Munich & Leipzig: Duncker & Humblot, 1920), 81.
7. Weber, "Konstitutionelle," AWG 7: 38–40.
8. Weber, "Der Beamte" (1910), AWG 8: 100. In the meeting he said that he knew his comments would not be well-received by the older generation because they had a predilection not only for the state but also for bureaucracy. Contribution to the discussion of "Die wirtschaftlichen Unternehmungen der Gemeinden" (1910), AWG 5: 426.
9. Ibid.; Weber, "Der Beamte," AWG 8: 99, 107.
10. Astrid Lange-Kirchheim has argued that Weber's *Neue Rundschau* essay was a direct inspiration for works such as "In the Penal Colony" and *The Trial* by his former student Franz Kafka. "Alfred Weber und Franz Kafka, in Demm, *Alfred Weber als Politiker*, 113–15, 121–25. This claim is supported by Walter Müller-Seidel, *Deportation des Menchen. Kafkas Erzählung "In der Strafkolonie" in europäische Kontext* (Stuttgart: Metzler, 1986).
11. Weber, "Der Beamte," AWG 8: 101–2.
12. Ibid., 116.
13. Jürgen Kocka, "White-Collar Employees and Industrial Society in Imperial Germany," in Iggers, *Social History of Politics*, 113–15.
14. Weber, "Der Beamte," AWG 8: 101–2, 107–8.
15. Ibid., 101.
16. Weber, Contribution to "Wirtschaftlichen Unternehmungen," AWG 5: 429–31, 435–36.
17. Weber, "Der Beamte," AWG 8: 100.
18. Weber, "Das Berufsschichsal der Indurtriearbeiter" (1912), AWG 8: 344–68; Weber, "Neuorientierung in der Sozialpolitik?" (1913), AWG 5: 475–85; Weber, "Die Bureaukratisierung und die gelbe Arbeiterbewegung" (1913), AWG 5: 459–74. Also see Kruse, *Soziologie und Gegenwartskrise*, 237–50.
19. Weber, "Neuorientierung," AWG 5: 475–84.

20. Weber, "Berufsschichsal," AWG 8: 349–50, 362–63; Weber, "Bureaukratisierung," AWG 5: 471–72.
21. Weber, "Neuorientierung," AWG 5: 482.
22. Weber, "Berufsschichsal," AWG 8: 364–68. Weber's colleague in the Association, Otto von Zwiedineck-Südenhorst, stated that choice of vocation was very important to Weber. "Zum Wirken von Max und Alfred Weber im Verein für Sozialpolitik: Erinnerungen und Eindrücke," in *Synopsis: Alfred Weber, 30.VII.1868–30.VII.1948*, ed. Edgar Salin (Heidelberg: Cambert Schneider, 1948), 784.
23. Weber, "Neuorientierung," AWG 5: 484.
24. Weber, "Bureaukratisierung," AWG 5: 459–62, 468, 471–74. This was an important qualification, because the most important unions were associated with the Social Democratic Party.
25. Weber, "Ber Beamte," AWG 8: 109–11; Contribution to "Wirtschaftlichen Unternehmungen," AWG 5: 427–29, 435–36.
26. Weber, "Der Beamte," AWG 8: 94–98, quote on 98.
27. Weber used this term in two ways. In this one, it was similar to its use by his brother in his lectures on vocation. He also applied to the bureaucracy, as noted above, to mean something like careerism.
28. Ibid., 113.
29. Weber, *Gedanken zur deutschen Sendung* (1915), AWG 7: 116–17.
30. Ibid., 121.
31. Ibid., 121–22.
32. Ibid., 133. Despite this somewhat aggressive language, Weber's position was at its heart defensive. Here, he sided with most German academics.
33. Ibid., 124–25.
34. A year later, the Battle of Jutland would render any strategy of annihilation of the British fleet meaningless.
35. Ibid., 134, 136–39, 159–62, 169–70.
36. Ibid., 133.
37. Henry Cord Meyer, *Mitteleuropa in German Thought and Action, 1815–1945* (The Hague: Springer, 1955), 87, 206.
38. Ibid., 95.
39. Quoted in Ibid., 136
40. Freidrich Naumann, *Mitteleuropa* (Berlin: Georg Reimer, 1916), 233; Meyer, *Mitteleuropa*, 233–35.
41. Naumann, *Mitteleuropa*, 100–1.
42. Meyer, *Mitteleuropa*, 233–37, 287–88. This discussion relies heavily on Meyer.
43. Meyer, *Mitteleuropa*, 290.
44. Demm, *Liberaler*, 162–69.
45. Weber, *Gedanken*, 7: 149.
46. Ibid., 171.
47. Demm, *Liberaler*, 202–8.
48. Weber, "Kontinental Verständigung" (1917), AWG 7: 202–8; Weber, "Aufstellung eines Friedensprogramms mit unseren Bundesgenossen" (1917), AWG 7: 187–93; Weber, "Das Selbstbestimmungsrecht der Völker und der Friede" (1918), AWG 7: 194–206.
49. Weber, "Kontinental Verständigung," AWG 7: 182–84.
50. Weber made the same point in an earlier memorandum. Weber, "Bemerkungen über die auswärtige Politik und die Kriegziele, in Demm, *Alfred Weber als*

Politiker, 171–77. At various times he took up the causes of Poland, Lithuania, and Finland. Demm, *Liberaler*, 192, 215, 225.
51. The cultural challenge of Russia was virtually ignored in this speech.
52. Weber, "Selbstbestimmungsrecht," AWG 7: 195–98. Weber also detected a utilitarian economic reason, the separation of Lorraine, with its rich mineral reserves, from Germany (194). Based on his theory of locations, Weber believed that Germany's economic future was guaranteed by its control of mineral holdings, especially coal. Weber, "Die Standortslehre und die Handelspolitik" (1911), AWG 6: 281–82.
53. Weber, "Selbstbestimmungsrecht," AWG 7: 196.
54. Ibid., 198–99.
55. This was an ironic assertion, since cartels played a much more important part in the German economy than they did in the British economy.
56. Ibid., 202–6, quote on 206.
57. Ibid., 197–98.

6 The Weimar Republic and the End of the Discursive Coalition

1. Peter Fritzsche, *Germans into Nazis* (Cambridge, MA: Harvard University Press, 1998), 72–73.
2. The left-liberal German Democratic Party was soon replaced by the right-liberal German People's Party.
3. Detlev J. K. Peukert, *The Weimar Republic: The Crisis of Classical Modernity* (New York: Hill & Wang, 1993); Dirk Käsler, *Die frühe deutsche Soziologie und ihre Entsteheungs-Milieus 1909 bis 1934* (Opladen: Westdeutscher Verlag, 1984), 204.
4. Bernd Widdig, *Culture and Inflation in Weimar Germany* (Berkeley & Los Angeles: University of California Press, 1999), 16. The following discussion of the inflation is greatly informed by Widdig's book.
5. David Frisby, *Fragments of Modernity* (Cambridge, MA: MIT Press, 1988).
6. Konrad Haenisch, "Sozialdemokratische Kulturpolitik," *Die Glocke* 4 (1918): 335–52; Haenisch, "Aus dem neuen Kultusministerium: Ein offener Brief an Professor Saenger," *Die neue Rundschau* 30 (1919): 17–27; Carl H. Becker, *Gedanken zur Hochschulreform* (Leipzig: Quelle & Meyer, 1919); Carl H. Becker. *Kulturpolitische Aufgaben des Reiches* (Leipzig: Quelle & Meyer, 1919); Carl H. Becker, *Vom Wesen der deutschen Universität* (Leipzig: Quelle & Meyer, 1925), 41. For an extended discussion of this controversy see Colin Loader and David Kettler, *Karl Mannheim's Sociology as Political Education* (New Brunswick, NJ: Transaction, 2002), 50–63, and Fritz K. Ringer, *The Decline of the German Mandarins* (Cambridge, MA: Harvard University Press, 1969), 69–70.
7. Becker, *Wesen der deutschen Universität*. 41
8. When the chair of sociology at the University of Frankfurt came open in 1929, Becker chose Alfred Weber's Heidelberg colleague, Emil Lederer, to fill it. Lederer decided to remain in Heidelberg, but it went to another Heidelberg academic, Karl Mannheim.
9. Georg von Below, "Soziologie als Lehrfach. Ein kritischer Beitrag zur Hochschulreform," *Schmollers Jahrbuch* 43 (1919): 59–110; Georg von Below, "Was ist 'Soziologie'? Eine Frage des Universitätsunterrichts," *Hochland*, 16 (1919): 550–55; Ernst Robert Curtius, *Deutscher Geist in Gefahr* (Stuttgart and Berlin: Deutsche Verlags-Anstalt, 1932).

10. Ernst Troeltsch, "Public Opinion in Germany: Before, During and After the War," *The Contemporary Review* 123 (1923): 582–83.
11. Ernst Troeltsch, "Die Krisis des Historismus," *Die neue Rundschau* 33 (1922): 581, 584, 589.
12. Charles R. Bambach, *Heidegger, Dilthey and the Crisis of Historicism* (Ithaca: Cornell University Press, 1995), 37–38, 41, 188; Ringer, *German Mandarins*, 245.
13. Except where noted, the following discussion rests primarily on three surveys: Peter C. Caldwell, *Popular Sovereignty and the Crisis of German Constitutional Law* (Durham, NC: Duke University Press, 1997); Peter M. R. Stirk, *Twentieth-Century German Political Thought* (Edinburgh: Edinburgh University Press, 2006); Chris Thornhill, *German Political Philosophy: The Metaphysics of Law* (London: Routledge, 2007); Arthur J. Jacobson and Bernhard Schlink, eds., *Weimar: A Jurispruence in Crisis* (Berkeley and Los Angeles: University of California Press, 2000).
14. Moritz Julius Bonn, *The Crisis of European Democracy* (New Haven: Yale University Press, 1925).
15. Ibid., 80–93.
16. Heidelberg academics tended to refer to themselves as "men of learning," rather than professors, to contrast themselves to specialists. They also separated themselves from "intellectuals," "aesthetes," and "litterateurs," all of which were associated with modernism and/or political radicalism. Christian Jansen, *Professoren und Politik. Politisches Denken und Handeln der Heidelberger Hochschullehrer 1914–1935* (Göttingen: Vandenhoeck & Ruprecht, 1992), 70–75.
17. Jansen, *Professoren*, 35–36, 86–88, 105–7. During the republic, all German universities continued to celebrate the founding of the empire without a similar ceremony for the republic. Often in their Founding Day speeches they explicitly criticized the republic or praised the empire in a way that implied criticism of the republic. At Freiburg, one speaker accused the "usurper" President Friedrich Ebert of "high treason." In Heidelberg, with one exception, such speeches did not occur. Ringer, *German Mandarins*, 214–16; Jansen, *Professoren*, 45–46.
18. Eberhard Demm, *Ein Liberaler in Kaiserreich und Republik. Der politische Weg Alfred Webers bis 1920* (Boppard, Germany: Harald Boldt, 1990), 262; Eberhard Demm, *Von der Weimarer Republik zur Bundesrepublik. Der politische Weg Alfred Webers 1920–1958* (Düsseldorf: Droste, 1999), 168, 174.
19. Demm, *Liberaler*, 256–82. In 1924, he was invited to join the newly formed Republican Party and wrote a draft programme (1924). The draft was never accepted, and the party quickly dissolved after its poor election showing. Demm, *Liberaler*, 291–92.
20. Quoted in Jansen, *Professoren*, 84.
21. Demm, Liberaler, 254–58; Herbert Döring, *Der Weimarer Kreis: Studien zum politischen Bewußtsein verfassungstreuer Hochschullehrer in der Weimarer Republik* (Meisenheim/Glan: Anton Hain, 1975), 204–5.
22. Marianne Weber, "Conviviality"; Gesa von Essen, "Max Weber und die Kunst der Geselligkeit" in *Heidelberg im Schnittpunkt*, ed. Treiber and Sauerland, 462–84.
23. Christian Jansen, "Das Institut der Außenseiter. Inneruniversitäre Spannungen und Öffentlichkeit," in Reinhard Blomert et al. (eds) *Heidelberger Sozial- und Staatswissenschaften: Das Institut für Sozial- und Staatswissenschaften zwischen 1918 und 1958* (Marburg, Germany: Metropolis, 1997), 29–30; Carl Zuckmayer, *Als wär's ein Stück von mir* (Vienna: Fischer, 1967), 301–2; Jansen *Professoren*, 35.

24. Blomert et al., *Heidelberger*, 12–13.
25. Jansen, "Institut," 31–38; Demm, *Von der Weimarer Republik zur Bundesrepublik. Der politische Weg Alfred Webers 1920–1958* (Dusseldorf, Germany, Droste, 1999), 116, 45, 61; Reinhard Blomert, *Intellektuelle im Aufbruch: Karl Mannheim, Alfred Weber, Norbert Elias und die Heidelberger Sozialwissenschaften der Zwischenkriegszeit* (Munich: Carl Hauser, 1999), 9.
26. Weber's student Carl J. Friedrich founded the Academic Exchange Service (today the DAAD), which had ties to the InSoSta. Exchange student Talcott Parsons wrote his dissertation at the Institute under Salin. Hans J. Lietzmann, "Carl Joachim Friedrich. Ein amerikanischer Politikwissenschaftler aus Heidelberg," in *Heidelberger*, ed. Blomert et al., 267–90.
27. Demm, *Weimarer Republik*, 46–48; Blomert et al., *Heidelberger*, 20.
28. Blomert, *Intellektuelle*, 18, 54, 108. Nevertheless, with funding from the Rockefeller Foundation, which hoped to promote inductive social research, the Institute became more research oriented.
29. Ibid., 271–80. Weber dated the actual writing of the book from 1931 to 1935. AWG 1: 52.
30. Originally the group, which contained sixty-four members, called itself the "Alliance of University Teachers True to the Constitution." The name "Weimar Circle" became official in 1931. Döring, *Weimarer Kreis*, 6–7, 257–60.
31. Ibid., v, 74–76, 83, 85, 92, 106.
32. Ibid., 14–17, 99, 209, 213, 215, 222. Although Weber was in many ways representative of the Weimar Circle, few of its members shared his enthusiasm for vitalism.
33. Guido Müller, "Der Publizist Max Clauss. Die Heidelberger Sozialwissenschaften und der 'Europäische Kulturbund (1924/5–1933)," in Blomert et al. (eds.), *Heidelberger*,.369–409; Demm, *Weimarer Republik*, 214–16.
34. Müller, "Max Clauss"; Demm, *Weimarer Republik*, 216–19.
35. Blomert et al., *Heidelberger*, 20.
36. Demm, *Weimarer Republik*, 223–31; Eberhard Demm, *Geist und Politik im 20. Jahrhundert. Gesammelte Aufsätze zu Alfred Weber* (Frankfurt/Main: Peter Lang, 2000), 309–23; Klaus-Rainer Brintzinger, "Die nationalsozialistische Gleichschaltung des InSoSta und die Staats- und Wirtschaftswissenschaftliche Facultät 1934–1945," in Blomert et al. (eds.), *Heidelberger,*, 55–81.

7 The Sociology of Culture in Weimar

1. The work, originally written in 1887 and significantly revised in 1911, appeared in a third, largely unchanged edition in 1920, and, in the next six years, went through four more editions.
2. Ernst Troeltsch, "Die Revolution in der Wissenschaft," in Troeltsch, *Gesammelte Schriften* (Tübingen, Germany: Mohr Siebeck, 1925) 4: 653.
3. Paul Forman documents the acceptance of vitalism by a significant number of physical scientists. See Forman, "Weimar Culture, Causality, and Quantum Theory, 1918–27: Adaptation by German Physicists and Mathematicians to a Hostile Intellectual Environment," *Historical Studies in the Physical Sciences* 3 (1971).
4. Another Weber associate, Count Hermann Keyserling, is also mentioned. See Troeltsch, "Revolution in der Wissenschaft," 664.
5. Ibid., 664–65.

6. H. Stuart Hughes, *Oswald Spengler: A Critical Estimate* (New York: Scribner, 1962), 1, 65; Bernd Widdig, *Culture and Inflation in Weimar Germany* (Berkeley & Los Angeles, 2001), 105; Forman, "Weimar Culture," 30. In 1922 Spengler's second volume appeared, and the following year the first volume was revised "but unchanged in its basic ideas." Hughes, *Oswald Spengler*, 1, 65, 89.
7. Ernst Cassirer, *The Myth of the State* (New Haven: Yale University Press, 1961), 289.
8. Ibid., 289–91.
9. Widdig, *Culture and Inflation*, 105; Georg Lukács, *The Destruction of Reason*, trans. Peter Palmer (London: Merlin, 1980), 462.
10. Hughes, *Oswald Spengler*, 89–94. Forman writes that the typical academic reaction was: "Of my discipline Spengler understands, of course, not the first thing, but aside from that the book is brilliant." Forman, "Weimar Culture," 30.
11. Spengler cites Goethe and Nietzsche as the two biggest influences on his work. Oswald Spengler, *The Decline of the West*, trans. Charles Francis Atkinson (New York: Knopf, 1932), 1: 49n. Bergson and Driesch are not cited, however, Spengler had clearly read the former. See Lukács, *Destruction*, 461, 464.
12. In fact, Spengler was more relativistic than most of the historicists in that he believed that natural science is not exempt from this relativism.
13. Nature was governed by the concept of causality.
14. Spengler, *Decline*, 2: 113–14, 165, 361–64.
15. Widdig, *Culture and Inflation*, 108.
16. Spengler, *Decline*, 2: 462, 506–7
17. Spengler, *Preußentum und Sozialismus* (Munich: Beck, 1922), 98.
18. While acknowledging that Spengler's book was a feat, Weber decried Spengler's use of his fame to publish reactionary brochures attacking democracy. Weber, "Oswald Spengler der Politiker," (1924), AWG 7: 448–52.
19. There are four examples of this type. "Prinzipielles zur Kultursoziologie: Gesellschaftsprozess, Zivilisationsprozess und Kulturbewegung," *Archiv für Sozialwissenschaft und Sozialpolitik* 47 (1920): 1–49, AWG 8: 147–86, at the beginning of the republic, is his most extensive explication of the discipline. His encyclopedia article toward the end of the republic, "Kultursoziologie," in an important 1931 compendium on sociology, largely abridges the earlier piece with some added terminology (AWG 8: 129–46). He offered a more specific and comparative account of his methodology in "Einleitung: Aufgabe und Methode," the introduction to a 1927 collection of previously published essays on culture and cultural politics (AWG 8: 35–59). These three essays were oriented primarily towards academics. In 1923, Weber published a short essay for a wider public in two venues—as "Kultursoziologie" in the literary magazine, *Der neue Merkur*, and as "Die Überwindung des Relativismus" in Hermann Keyserling's journal, *Der Leuchter*. It was reprinted in his 1927 collection as "Kultursoziologie und Sinndeutung der Geschichte," AWG 8: 76–82. There are other texts that examine culture in relation to politics. They will be treated more extensively in the next chapter. "Prinzipielles" will be cited in both its original and AWG editions, due to some minor differences.
20. "Kultursoziologische Versuche: Die alte Ägypten und Babylonien" (1926), AWG 8: 203–52; and a 1935 book that he considered his most important work, *Kulturgeschichte als Kultursoziologie* AWG 1.
21. In 1927 Weber was freed from his obligation to teach national economics and began to offer lectures and seminars on the sociology of culture that covered both of these aspects, although the second was emphasized.

22. Weber, "Prinzipielles," [1] 147. The original *Archiv* version is in brackets.
23. I have chosen to translate *"Kreis"* literally as "circle," rather than "sphere," which Weltner and Hirshman choose, because it contains both a spatial and temporal aspect and because Weber uses *"Sphäre"* in another way. It also resonates with the different intellectual circles of Heidelberg discussed above. By 1931, he no longer used the term "historical circle."
24. Salomon Wald, *Geschichte und Gegenwart im Denken Alfred Webers* (Zurich: Polygraphischer, 1964), 21.
25. The more general and inclusive "stream" (*Strom*) is replaced in this essay with "current" (*Strömung*).
26. Weber, "Kultursoziologie," AWG 8: 134.
27. Volker Kruse, *Soziologie und "Gegenwartskrise." Die Zeitdiagnosen Franz Oppenheimers und Alfred Webers* (Wiesbaden, Germany: Deutscher Universitäts Verlag, 1990), 263.
28. Roland Eckert, *Geschichtstheorie*, 30, 72. My approach is also that of Eckert using somewhat different language.
29. Weber, "Prinzipielles," [11–13], 8:155–57.
30. For the general use of "spirit" at the time, see the well-known encyclopedia, *Der grosse Brockhaus*, fifteenth edition (Leipzig, Germany: Brockhaus, 1928–1935), VII: 98–99.
31. Weber, "Prinzipielles," [4–5]. In the later version he added biology, AWG 8: 149–50.
32. Alexander von Schelting, "Zum Streit um die Wissenssoziologie I. Die Wissenssoziologie und die kultursoziologische Kategorien Alfred Webers," *Archiv für Sozialwissenschaft und Sozialpolitik* 62 (1929): 20–22.
33. Weber, "Prinzipielles," [14–21], 8: 157–64.
34. I have translated this term literally despite its religious ring, because there is no better equivalent. The problem arises with the adjectival form. Here I have chosen to use "psychic" rather than the obsolete "soulish."
35. Peter Gay, *Weimar Culture* (New York: Harper & Row, 1968), 70–101; also see Kruse, *Gegenwartskrise*, 263. See *Der grosse Brockhaus*, XVII, 214–16, for the ambiguity of this term.
36. Weber, "Prinzipielles," 29; Weber, "Kultursoziologie und Sinndeutung der Geschichte," in *Ideen*, 49; Weber, *Die Krise des modernen Staatsgedanken in Europa* (1925), AWG 7: 45.
37. Weber, "Prinzipien," 25–26. This depiction is very much in accord with his replacing the more inclusive and defined "stream" (*Strom*) with "current" (*Strömung*) in this article.
38. Weber, "Sinndeutung," AWG 8: 77–78.
39. Weber, "Aufgabe," AWG 8: 52–54. He also dealt with Ernst Troeltsch and Max Scheler.
40. Weber, "Aufgabe," AWG 8: 54.
41. There is a large and rich literature on Max Weber's methodology. In delineating its application, Stephen Kalberg, *Max Weber's Comparative-Historical Sociology* (Chicago: University of Chicago Press, 1994), is especially useful.
42. Max Weber, "'Objectivity' in Social Science and Sociology," in Max Weber, *The Methodology of the Social Sciences*, trans. Edard A. Shils and Henry A. Finch (New York: Free Press, 1949), 80.
43. Max Weber, "Objectivity', 90 (original emphases).
44. Max Weber, *Economy and Society*, ed. Gunther Roth and Claus Wittich (Berkeley & Los Angeles: University of California Press, 1978), 24–26.

45. For example, Weber tells of attending a Baptist baptism in a cold mountain stream in North Carolina. When a Mr. X was baptized, Weber's cousin explained that Mr. X planned to open a bank and needed credit. His admittance to the Baptist congregation was of importance to his Baptist customers, but more so to his non-Baptist customers, because it certified that his moral and business conduct had passed an on-going set of inquiries by the congregation. Max Weber, "'Churches' and 'Sects,'" 8.
46. Max Weber, *Economy and Society*, 2–27.
47. Ibid., 40–41. Weber acknowledges that these two types are based on Ferdinand Tönnies's earlier pair, community (*Gemeinschaft*) and society (*Gesellschaft*). His modification allows him to escape the rigidity of Tönnies's dichotomy for something more resembling a spectrum as is in keeping with his concept of the ideal type.
48. Max Weber, *Economy and Society*, 53–54, 215–16.
49. Weber, "Aufgabe," AWG 8: 44.
50. Ibid., 37–38. Throughout this essay, Weber alternates the term "constellation" with "physiognomy" and "life-aggregation," the latter two having a more vitalistic ring. "Constellation" became more prominent in his writings after 1945. See Willi, "Konstellationssoziologie."
51. Weber, "Aufgabe," AWG 8: 44–47.
52. Weber, "Alte Ägypten," AWG 8: 203–52.
53. Ibid., 203–5.
54. Ibid., 211–12, 217–19.
55. Ibid., 227–36.
56. Ibid., 248–51.
57. Weber, "Kultursoziologie," AWG 8: 138.
58. See Kruse, *Soziologie und Gegenwartskrise*, 268–74.
59. Weber, "Kultursoziologie," AWG 8: 134–35.
60. See Eckert, *Geschichtstheorie*, 125–32; Matti Luoma, *Die Drei Sphären der Geschichte. Systematische Darstellung und Versuch einer kritischen Analyse der kultursoziologischen inneren Strukturlehre der Geschichte von Alfred Weber* (Helsinki, Finland: Societas Scientiarum Fennica, 1959), 35; Victor Willi, "Bemerkungen zur sogenanten Konstellationssoziologie Alfred Webers," 655–56.
61. Weber, *Kulturgeschichte als Kultursoziologie*, AWG 1: 61. Page numbers in parentheses refer to this version, which incorporates both the 1935 and the 1950 editions. My discussion applies primarily to the former.
62. Weber, "Kultursoziologie," AWG 8: 140–46; Eckert, *Geschichtstheorie*, 109.
63. The least political, *The Tragic and History* [*Das Tragische und die Geschichte*], AWG 2, appeared in 1943; *Farewell to the Previous History* [*Abschied von der bisherigen Geschichte*], AWG 3, was written at the end of the war and appeared in 1946. It was translated the next year as *Farewell to European History. The Third or the Fourth Person* [*Der dritte oder der vierte Mensch*], AWG 3, appeared in 1953. The revised and expanded edition of *Kulturgeschichte als Kultursoziologie* AWG 1, was published in 1950.
64. Luoma, *Drei Sphären der Geschichte*, 101.
65. *Kulturgeschichte*, AWG 1: 73–80, quotation is on 80.
66. Ibid., 41–44, 84–92.
67. Ibid., 137. Included in the East were the Islamic empire, Byzantium, and Russia. The West was limited to Europe west of Russia.
68. Ibid., 328–41, 390–404, 422–26.
69. Ibid., 439–50, 523.

8 Cultural Politics in Weimar

1. Max Weber, *Economy and Society*, ed. Guenther Roth & Claus Wittich (Berkeley & Los Angeles: University of California Press, 1978), 1381–1469; Max Weber, *From Max Weber*, ed. Hans H. Gerth and C. Wright Mills (New York: Oxford University Press, 1946), 77–156. For the background to these concerns, see especially Wolfgang J. Mommsen, *Max Weber and German Politics 1890–1920*, trans. Michael S. Steinberg (Chicago: University of Chicago Press, 1984), 137–282.
2. Weber, *Economy and Society*, 1401–2.
3. The German term translated here as "objectified spirit" (*geronnener Geist*) conveys the sense of something that once flowed becoming congealed. We have seen this same imagery in Alfred's writings.
4. Weber, *Economy and Society*, 1402, translation modified; *Max Weber Gesamtausgabe*, ed. Gangolf Hübinger and Wolfgang J. Mommsen (Tübingen: Mohr Siebeck, 1984) I/15: 464–65.
5. Weber, *Economy and Society*, 925–26; David Beetham, *Max Weber and the Theory of Modern Politics* (Cambridge, UK: Polity, 1985), 121–31.
6. Max Weber, "Politics as a Vocation," in *From Max Weber*, ed. Hans Gerth & C. Wright Mills (Oxford: Oxford University Press, 1946), 95, 110–17; Weber, *Economy and Society*, 2047.
7. Weber, *Economy and Society*, 250–60; Weber, "Politics as a Vocation," 106–8; Beetham, *Max Weber*, 231.
8. Weber, "Die Bedeutung der geistigen Führer in Deutschland" (1918), AWG 7: 359–60.
9. Walter Struve labels both brothers as "liberal elitists [who]...subordinated a class analysis of society and a materialist view of history to an approach that recognized elites as the ultimate agencies of historical change." He uses Alfred's above-cited essay as an example of this type. Struve, *Elites against Democracy* (Princeton: Princeton University Press, 1973), 6–7. Here, Struve exaggerates the brothers' antidemocratic views.
10. Weber, "Bedeutung," AWG 7: 361.
11. Ibid., 349, 356–61.
12. Ibid., 348, 360–63.
13. Ibib., 354–55, 359, 361–66.
14. Ibid., 361, 365–67.
15. Weber, "Das Programm der Regierung," AWG 6, 230–32.
16. Weber, *Die Krise des modernen Staatsgedankens in Europa* (1925), AWG 7: 317, 321.
17. Ibid., 234–35, 253, 274–75.
18. Ibid., 236.
19. Ibid., 237–38, 252–67, 292–93.
20. Ibid., 274–77, 299–300.
21. Ibid., 279–83, 300–1, 309–10.
22. Ibid., 278, 282, 310–11, 313–14; Weber, "Geist und Politik," AWG 7: 377.
23. Weber, *Krise*, AWG 7: 321.
24. Ibid., 317.
25. Weber, "Geist und Politik," AWG 7: 386–87.
26. Bernd Widdig, *Culture and Inflation in Weimar Germany* (Berkeley & Loa Angeles: University of California Press, 2001), 182–95.
27. Weber, *Die Not der geistigen Arbeiter* (1923), AWG 7: 607–13.

28. Ibid., 603–8.
29. Ibid., 613–16.
30. Ibid., 607–8, 621–24, 627.
31. Widdig, *Culture and Inflation*, 183–85.
32. Ibid., 185–86.
33. Bourdieu, *Homo Academicus*, 112–18.
34. Weber, *Die Not der Geistigen Arbeiter*, AWG 7, 608, 627.
35. Mannheim, *Structures of Thinking*, ed. David Kettler et al., trans. Jeremy J. Shapiro and Shierry Weber Nicholsen (London: Routledge, 1982), 266–69.
36. Colin Loader, *The Intellectual Development of Karl Mannheim* (Cambridge, UK: Cambridge University Press, 1985), 84–94.
37. Emil Lederer, "Aufgaben einer Kultursoziologie," in Melchior Palyi (ed.), *Hauptprobleme der Soziiologie. Erinnerungsgabe für Max Weber*, vol. 2 (Munich and Leipzig: Duncker & Humblot, 1923), 149–51.
38. Ibid., 150–57.
39. Ibid., 161, 167–71. Lawrence Scaff focuses on this last point as an anticipation of recent views on modernity. Scaff, "Modernity and the tasks of a sociology of culture," 89–90.
40. Mannheim repeated this distinction in Karl Mannheim, *Ideology and Utopia*, trans. Louis Wirth and Edward Shils (New York: Harcourt, 1936), where he explicitly referred to Weber's categories (179).
41. Mannheim, "Competition as a Cultural Phenomenon," in *Knowledge and Politics*, ed. Volker Meja and Nico Stehr (London: Routledge, 1990), 56–58, 67–70. For an account of the background of this clash, see David Frisby, *The Alienated Mind* (London: Heinemann, 1983), 185–88, and Norbert Elias, *Reflections on a Life*, trans. Edmund Jephcott (Oxford: Polity, 1994), 109–20.
42. Alfred Weber, contribution to "Discussion of Karl Mannheim's 'Competition' paper at the Sixth Congress of German Sociologists," in Volker Meja and Nico Stehr (eds.), *Knowledge and Politics: The Sociology of Knowledge Dispute* (London: Routledge, 1990), 89–90.
43. Weber was not alone in his concern about Mannheim's pluralistic approach. Ernst Robert Curtius, a colleague of the two men at Heidelberg with a more conservative organic orientation than Weber, attacked Mannheim in print, in a critical review of *Ideology and Utopia* and in a 1932 book, *German Spirit in Danger*. See Curtius, "Sociology and Its Limits," in Meja and Stehr (eds.), *Knowledge and Politics*, 113–20; Ernst Robert Curtius, *Deutscher Geist in Gefahr* (Stuttgart & Berlin: Deutsche Verlags-Anstalt, 1932). Curtius also critiqued Max Weber's "Science as a Vocation," to which he attributed the same kind of pluralism as that of Mannheim. The disagreement between Alfred and Mannheim remained much more civil. On the conflict between Curtius and Mannheim, see Dirk Hoeges, *Kontroverse am Abgrund. Ernst Robert Curtius und Karl Mannheim. Intellektuelle und "freischwebende Intelligenz" in der Weimarer Republik* (Frankfurt/Main: Fischer, 1994); and Loader, *Intellectual Development*, 116–18.
44. Karl Mannheim, *Sociology as Political Education*, trans. and ed. David Kettler and Colin Loader (New Brunswick, NJ: Transaction, 2001) 112–13, 122.
45. Ibid., 113, 117–20.
46. Mannheim, *Ideology and Utopia*, 153–59, quotation on 159, translation modified.

47. The section of *Ideology and Utopia* dealing with the intelligentsia ends with a discussion of Max Weber (1936: 190–91).
48. Weber connected these two aspects briefly in "Mythologie oder Wirklichkeit?" (1927), AWG 7: 549–52.
49. Eberhard Demm, Ein *Liberaler in Kaiserreich und Republik. Der politische Weg Alfred Webers bis 1920* (Boppard, Germany: Harald Boldt, 1990), 209, 294.
50. As he had done earlier, Weber relegated France to a lesser role in this scenario. France was the inheritor of the mantel of Rome, that is, of an earlier form of imperialism. But that imperialism had been superseded by the new Anglo-American model and, so, the French-German rivalry had been rendered obsolete. France had become a junior partner to the new Anglo-American world, whose main purpose was to weaken Europe's sense of its own identity Because France had no ties to eastern Europe it could not participate as a formulator of the new European aspiration. Weber, *Deutschland und die europäische Kulturkrise* (1924), AWG 7: 482–84, 488–89; Weber, *Krise,* AWG 7: 161–62, 169.
51. Weber wrote: "The League of Nations in its essence is based upon the denial of Europe as a still existing historical body especially with a configured and balanced system of states, with a particular, still existent real and spiritual communal basis." Weber, *Krise,* AWG 7: 339–41, 344.
52. Weber, "Deutschlands finanzielle Leistungsfähigkeit jetzt und künftig" (1922), AWG 5: 523–24.
53. Weber, "Eine Rede des Professors Alfred Weber" (1926), AWG 7: 536–38.
54. Weber, *Deutschland und Kulturkrise,* AWG 7: 495–97.
55. Ibid., 480; Weber, "Amerika—Europa. Zu Arthur Feilers Amerika-Buch" (1926), AWG 8: 392–96.
56. Weber, *Deutschland und Europa 1848 und Heute* (1923), AWG 7: 503, 509–11, 514–15; Weber, *Deutschland und Kulturkrise,* AWG 7: 474–78; Weber, *Krise,* AWG 7: 293, 299–300.
57. Weber, *Deutschland und Kulturkrise,* AWG 7: 477–80; Weber, *Krise,* AWG 7: 340–42. Weber criticized his old friend, Czech President Thomas Masaryk for depicting Germany as essentially authoritarian. Weber, "Tschechische geschichtliche Mission und deutscher Geist. Offener Brief an den Präsidenten T. G. Masaryk" (1926) AWG 7: 517–21.
58. Weber, "Rede beim Schlussbankett der internationalen Jahresversammlung in Mailand" (1926) AWG 7: 533–35; Weber, "Eine Rede des Professors Alfred Weber" (1926), AWG 7: 536–38; Weber, "Der Deutsche im geistigen Europa" (1927), AWG 7: 539–48; Weber, "Zur Krise des europäischen Menschen" (1932), AWG 8: 418–22.
59. He had this doubt about political parties through much of the republic. See for example Weber, "Die Reichstagswahlen—und nun was? Ein Brief" (1924), AWG 7: 446–47; and Weber, "Verjüngung" (1924), AWG 7: 453–54. However, the emphasis was greater with the parliamentary gridlock at the end of the republic.
60. Weber, "Eigene Gedanke Alfred Webers über die Demokratie" (1929). AWG 7: 460–61.
61. Ibid. Here Weber was using the language of the Weimar youth movement, whose predominant form was the *Bund.*
62. Weber, *Das Ende der Demokratie? Ein Vortrag* (1931), AWG 7: 396–400.
63. Ibid., 402.

64. Weber, "Eigene Gedanke," AWG 7: 460–61; Weber, *Ende der Demokratie*, AWG 7: 389–90; Weber, "Wie das deutsche Volk fühlt!" (1931), AWG 7: 534–35.

9 Epilogue: Alfred Weber After 1933

1. Alfred Weber, "Soziologie" (1960), AWG 8: 752.
2. Ibid., 752–57.
3. Alfred Weber, *Farewell to European History or the Conquest of Nihilism*, trans. R. F. C. Hull (New Haven, CT: Yale University Press, 1948), 172.
4. Klaus von Beyme, Contribution to Eberhard Demm (ed.), *Alfred Weber zum Gedächtnis* (Frankfurt/Main: Peter Lang, 2000), 219–20.
5. Eberhard Demm, *Von der Weimarer Republik zur Bundesrepublik. Der politische Weg Alfred Webers 1920–1958* (Düsseldorf, Germany: Droste, 1999), 315–20; Eberhard Demm, "Geist und Politik," in Hans G. Nutzinger (ed.), *Zwischen Nationalökonomie und Universalgeschichte. Alfred Webers Entwurf einer umfassenden Sozialwissenschaft in heutiger Sicht* (Marburg, Germany: Metropolis, 1995),, 62–63; Steven P. Remy, *The Heidelberg Myth*, (Cambridge, MA: Harvard University Press, 2002), 220. Weber's writings on these issues are collected in AWG 9, for example, 17–69, 125–34, 331–33.
6. Demm, *Weimarer Republik*, 371–82; Demm, "Geist und Politik," 77–91.
7. Demm, *Weimarer Republik*, 412–16, 323.
8. Remy, *Heidelberg Myth* 16–22.
9. Remy, *Heidelberg Myth*, 117. Remy bases the first of these elements on a somewhat skewed reading of Christian Jansen's study of the Heidelberg professoriate to make them appear more conservative than they were. Jansen acknowledges that the entire "political spectrum" existed among the faculty, but he states that the university "was considered the stronghold of academic political liberalism until 1933. It and the newly established Frankfurt University were the only ones in which until 1930 faculty dominated who were neutral or supportive regarding the republic." Christian Jansen, *Professoren und Politik. Politisches Denken und Handeln der Heidelberger Hochschullehrer 1914–1935*. (Göttingen, Germany: Vandenhoeck & Ruprecht, 1992), 14. With regard to the final point, Demm points out that almost 42 percent of the faculty and 66 percent of the full professors were removed, although ten of the latter were allowed to return via a local civilian hearing board (*Spruchkammer*), Demm, *Weimarer Republik*, 328.
10. See, for example, his exchange of letters with Hans Frank. AWG 10: 306.
11. Remy, *Heidelberg Myth*, 110–13; Demm, *Weimarer Republik*, 238–41.
12. This principled stand was not without a cost. Two instructors who had been restored to their positions despite indirect but clear connections to the Nazis, the historian Johannes Kühn and the political economist Helmut Meinhold, took revenge on Weber by sabotaging the academic careers of a number of his students. Remy, *Heidelberg Myth*, 130, 138, 144, 166, 227; Demm, *Weimarer Republik*, 323–34, 342–48.
13. Weber, "Haben wir Deutschhen nach 1945 versagt?" (1949), AWG 9: 91. Three years earlier, he had declared the German people complicit in the crimes of the Nazis. "Zu dem warnungen F. W. Forsters in der deutschen Frage" (1946), AWG 9: 71–72.
14. Kurt Lenk, *Marx in der Wissenssoziologie* (Neuwied, Germany: Luchterhand, 1972), 9–41.

Appendix 1: A Note on the Translation

1. Hans-Ulrich Wehler, "Reiter- und Immerweitervolker. Alfred Weber hat den Aufgalopp zur modernen Kulturgeschichte verpaßt," *Frankfurt Allgemeine Zeitung* 238 (October 14, 1997), L36.
2. Walter Benjamin, "The Task of the Translator," in *Theories of Translation: An Anthology of Essays from Dryden to Derrida*, ed. Rainer Schulte and John Biguenet (Chicago: University of Chicago Press, 1992), 71.

Appendix 2: Fundamentals of Cultural Sociology: Social Process, Civilizational Process and Cultural Movement" (1920)

1. "Prinzipielles zur Kultursoziologie (Gesellschaftsprozeß, Zivilisationsprozeß und Kulturbewegung), *Archiv für Sozialwissenschaft und Sozialpolitik* (1920), vol. 47, 1–49. Modified 1951 version in AWG 8: 147–86.
2. Trans. note: These lectures were actually given the following year. See Eberhard Demm, Ein *Liberaler in Kaiserreich und Republik. Der politische Weg Alfred Webers bis 1920 (Boppard, Germany: Harald Boldt, 1990)*, 138–39.
3. As does, for example, [Oswald] Spengler's book on *The Decline of the West* [trans. Charles Francis Atkinson (New York: Knopf, 1932)].
4. Trans. note: This paragraph was eliminated from the 1951 version of the essay included in AWG, 8.
5. It should be noted that this concept is not limited to Spengler, but rather implicitly or explicitly is the basis of all recent historical writing. Just as the "youth" or "old age" of bodies has obviously been an ingredient of this viewpoint *for a long time*.
6. Despite the brilliant personal contributions of Jacob Burkhardt, above all, and a few others.
7. This essay does not fail to recognize comprehensive treatments, such as those of Max Weber and Troeltsch in the sphere of the history of religion, or the somewhat "spontaneous" attempts that occur in numerous recent treatises from various cultural fields.
8. Trans. note: In 1951, this is rendered as *"psychic*-spiritual."
9. Clearly, it is a matter here of marginal issues of the materialistic conception of history. The latter's formulation of "interests" does not lead to the clarification of the decisive categories of intuition.
10. Despite the many points in common between the above and the formulations of Max Weber in his essays on the sociology of religion, the latter proceed from a somewhat different viewpoint. Unfortunately, here it is not possible to differentiate between the two positions.
11. This is said without accepting or rejecting the Bergsonian intuitions *"etiologically"* or *epistemologically* within the sphere of their *philosophical* formulation.
12. Hegel's protest against the overvaluation of the understanding (*Verstand*) does not change the fact of his entanglement in this attitude through his all-powerful concept of reason.
13. Trans. note: reference here to Johann Plenge's book, *Marx and Hegel* (Tübingen, Germany: Laupp, 1911).

Index

academia, *see* professors
Academic Exchange Service, 233
Academy for Labor, Frankfurt, 110
Action Group for Democracy and Free Socialism, 160
Adenauer, Konrad, 160–1
aggregation, 7, 74, 140–1, 182, 189, 192–4, *also see* life-aggregation
Akhenaten, 127–8
Alexander, Jeffrey, 223, 227
America, United States of, 87, 90, 94, 99, 135, 152–4
analysis, 63, 66, 67, 74, 75, 78, 119, 123, 127, 142, 149, 166, 203, 226, 237
Anschütz, Gerhard, 106–7, 111
apparatus, 67–8, 84–5, 87–8, 102, 125, 128, 138–9, 173, 175–6, 178, 180, 187
"Appeal to the World of Culture," 58
Arabian circle, culture, 117, 166, 176–8, 180, 182, 199, 202
Archive for Social Science and Social Policy, 29, 77
aristocracy, *see* elite
Aron, Raymond, 212
Ascheim, Stephen E., 217
aspiration, *see* will
Austrian School of Marginal Utility, 6, 21, 24, 31, 39–41, 45, 47–8, 146, 219

Babylonia, ancient, 8, 120, 127, 129–30, 176–7, 182
Baden-Badener Society, 221
Baehr, Peter, 223
Bambach, Charles R., 232
Barkin, Kenneth D., 216
Becker, Carl H., 101–2, 231
Beetham, David, 223, 237
Below, Georg von, 102, 231
Benjamin, Walter 163, 241
Bergson, Henri, 39, 57, 64–7, 73, 120, 129, 225, 234, 241

Bergstraesser, Arnold, 111
Berlin, 6, 47, 49, 103, 113
    University of, 21, 31
Bethmann-Hollweg, Theobald von, 99
Beyme, Klaus von, 159, 240
Bildung, see cultivation
binary, 6–7, 16–18, 20, 23–5, 27–9, 32–3, 38, 43, 45, 58, 63, 72–3, 75–6, 82, 85, 117–21, 124, 131, 133, 136, 139, 141, 147, 151–3, 157, 172, 216–17, 226–8, 236
Bismarck, Otto von, 26, 83, 135
Blackbourn, David, 216
Bleicher, Josef, 212
Bloch, Ernst, 57
Blomert, Reinhard, 211, 223, 232, 233
body (*Körper*) 43–4, 69, 93, 121–2, 131, 142, 167, 169, 171, 174,
    also see historical body
Böhme, Klaus, 224
Bonn, Mortiz Julius, 104, 232
Bostaph, Samuel, 215
Bourdieu, Pierre, 2–3, 143, 145, 148, 212, 238
Böventer, Edwin von, 220
Brentano, Lujo, 216, 217
Breuer, Stefan, 221
Brintzinger, Klaus-Rainer, 233
Britain, 58–9, 90–1, 93–5, 108, 153–4
Brockhaus, Der grosse, 235
Brod, Max, 39
Bruch, Rüdiger vom, 25, 214, 216
Brüning, Heinrich, 156
Buddhism, 182–3, 191, 226
bureaucracy, 2, 5, 8, 13, 16–20, 22–3, 26–8, 33, 41, 44–5, 55, 64, 68, 76, 81–8, 103, 128, 133–7, 139, 142, 151, 155, 157–9, 193, 223, 225, 228, 229, 230, *also see* state, bureaucratic
Burkhardt, Jacob, 62–3, 241
Buselmeier, Michael, 222

243

244 Index

Caesarism, 119, 136, 169
Caldwell, Peter C., 232
Campbell, Joan, 224
capitalism, 14, 26, 28, 34, 36–9, 40, 44, 55, 61, 68, 76, 79, 84–6, 90, 100, 109, 131, 134, 139–41, 145, 148, 161, 192, 194, 217, 219, 226
Cassirer, Ernst, 116–17, 234
center, 6, 7, 9, 15, 51, 54, 64, 67, 69, 70, 73–5, 77, 80–1, 84, 88, 92, 95, 107, 125, 128–9, 138, 140, 144, 151–4, 158, 168, 179, 190, 192, 198, 203
Center Party, 99–100
Central Europe, *see* Mitteleuropa
Chatman, Seymour, 213
China, 117, 131, 166, 169, 172, 176–7, 182, 193, 202
Christianity, 23, 80, 182–4, 191, 226
"civil peace," *see* "Ideas of 1914"
civil society, 17–19, 23, 28, 31, 35, 38, 43, 47, 58, 70, 76, 81, 101–2, 104–7, 119, 141–2, 146, 157, 216, *also see* society
civilization, 6–7, 17, 43, 63, 73–5, 78, 80, 82, 90, 94–5, 117–18, 129, 133–5, 137, 139–40, 146, 155, 159, 184, 186–7, 194–5, 197, 199–200, 202–3, 217, 220, 226
civilizational process, 6–7, 68, 79, 122–4, 127–8, 131, 153, 166, 173–81, 184, 188–9, 192–3, 196, 198, 200
Classical circle, culture, 117, 166, 172, 176–84, 191, 193, 199, 202
Clauss, Max, 112
clothing, ready-made (*Konfektion*), 32–8, 217
Cold War, 160–1
College for Politics, Berlin, 110
Committee of Thirteen, 162
communicative knowledge, 147
community (*Gemeinschaft*), 16–17, 20, 22, 52, 55–6, 58, 60, 69, 70–2, 86, 105, 107, 114, 118, 120, 124, 147, 152, 157, 168, 216, 236
Comte, Auguste, 74, 196
Condorcet, Marquis de, 196
configuration (*Gestaltung*), 6, 7, 18, 41, 43, 60, 67, 86, 94, 120, 123–5, 129, 135, 138, 141, 166, 168–71, 173, 188–90, 192–5, 198–9, 212

Confino, Alon, 49, 220, 221
conjunctive knowledge, 147
Conrad, Johannes, 215
constellation, 18, 66, 71, 77, 86, 127, 130, 132, 151, 227, 236
constitutional law, 103–6
Continental Understanding, 93
Copernicus, Nicholas, 121, 175, 189
core, *see* center
corporeal, *see* body
crisis, 4, 15, 17–18, 100, 103–4, 107, 134, 141, 143, 158
 of classical modernity, 100, 116
 of culture, 2, 7–8, 25, 45, 55, 64, 67, 73, 75, 88, 124, 131–2, 134, 143
cultivated culture, 147
cultivation, 5, 15, 19, 22, 51, 61, 63, 88, 102, 107–8, 116, 119, 129, 150, 157, 159
cultural capital, 4, 5, 15, 20, 47, 51, 101, 115, 134, 143, 145, 146
cultural community, 25, 94, 134, 138, 156
cultural currents, 79, 171, 185, 191, 204
cultural history, 8, 9, 120, 166–7, 204
Cultural League for the Democratic Renewal of Germany, 160
cultural movement, 8, 45, 74, 78, 116, 121, 123–4, 127, 166, 173–4, 176, 178, 181–2, 184–5, 188, 191–3, 195–6, 200–4
cultural question, 40, 54, 60–1, 64, 67–73, 77, 86, 186, 199, 203
cultural re-aggregation, 191
Cultural Sociological Seminar, 113, 146
cultural sociology, *see* sociology, of culture
cultural symbols, 7, 123, 180, 182, 199, 200–1
culture, 6, 17, 27, 31, 43, 62, 64, 69, 73–6, 78–80, 90, 95, 101, 108, 117–18, 122–3, 129, 133, 142, 144, 146, 150, 152–3, 159, 168, 171, 186–8, 190, 194, 197–203, 217, 220, 226
 creativity of, 147, 149, 184, 190–1
 emanation of, 168, 184, 190
Curtius, Ernst Robert, 102, 231, 238
Curtius, Ludwig, 50, 53

Dahlmann, Friedrich Christoph, 20
Demm, Eberhard 2, 10, 211–12, 213, 217, 218, 219, 222, 223, 224, 225, 227, 228, 230, 232, 233, 239, 240, 241
democracy, 94, 104, 116, 119, 137, 139, 141–2, 154–5, 159–60, *also see* leader democracy
Dempsey, Bernard, 219
denazification, 161–2
Dericum, Christa, 223
destiny, 21, 72, 85–7, 118, 120, 123, 138, 158, 166–71, 176, 199, 200, 202–3
diagnosis, 16–17, 45, 88, 117, 120
dichotomy, *see* binary
dictatorship, 105
Dilthey, Wilhelm, 57, 225
discursive coalition, 3–6, 13–18, 20–9, 31–2, 45, 47–9, 53–6, 58, 62–76, 81–5, 87–8, 99, 100–1, 103–4, 107–9, 111, 115–19, 133–4, 137, 139, 151, 157–8, 162, 228
Döring, Herbert, 232, 233
Driesch, Hans, 57, 64–6, 73, 120, 225, 234
dualism, *see* binary

Eastern Europe (*Osteuropa*), 92, 152
Ebert, Friedrich, 232
Eckardt, Hans von, 113
Eckert, Roland, 218, 227, 235, 236
Egypt, ancient, 8, 117, 120, 127, 129–31, 134, 160, 166, 169, 176–8, 182, 193
Elias, Norbert, 2, 110, 113, 146, 160, 212, 238
elite, 5, 10, 17, 19, 26, 28, 37, 48, 51–2, 54, 63, 81, 83, 101–3, 108, 111, 118–19, 136, 141, 159, 237
   spiritual, cultural, 9, 15, 63, 82, 112, 137, 139–40, 142–4, 146–7, 151–2, 157–8, 162
empiricism, 25, 153, 226
Enlightenment, 6, 24, 67, 121, 124, 140
entelechy, 65–6, 120, 123, 129, 227
Eschmann, Ernst Wilhelm, 111
Essen, Gesa von, 232
Eßlinger, Hans Ulrich, 211
eternal conversation, 50, 52, 55, 101

ethic of conviction, 135
ethic of responsibility, 52, 135, 138, 160
Euclid, 121, 175, 180
Europäische Revue, 112
European Cultural League, 110–11

fascism, 101, 111, 155–6
Fatherland Party, 92, 99
Federal Republic of Germany, 9, 162
Fichte, Johann Gottlieb, 196–7
field, 3, 4, 5, 14–15, 42, 45, 47–9, 55, 77, 145, 148, 167, 215, 241
foreign policy, 8, 89, 91, 152–4
Forman, Paul, 233, 234
fourth man, 132, 159
Franger, Wilhelm, 221
Free German Youth, 57, 69–70, *also see* youth movement
free socialism, 161
Freyhofer, Horst H., 226
Friedrich, Carl J., 233
Frisby, David, 231, 238
Fritzsche, Peter, 220, 224, 231

Gay, Peter, 123, 235
Gemeinschaft, *see* community
George, Stefan, 55–7, 124, 222, 223
   circle of, 55–7, 110, 116, 163, 224, 225
German Democratic Party (DDP), 100, 109, 111, 156, 231
German National People's Party (DNVP), 113, 152
German People's Party (DVP), 111, 231
German Society of Voters, 160
German Sociological Association, 29, 62, 73, 110, 120
German-Romance culture, 166, 179, 183
Gerth, Hans, 2, 110, 113, 146
Gesellschaft, *see* society
Geuss, Raymond, 17, 214, 216
Gierke, Otto von, 104, 140
Gilbert, Edmund W., 221
Giovanni, Norbert, 211
Gneist, Rudolf, 215
Goethe, Johann Wolfgang von, 234
Goldman, Harvey, 223
Gorges, Irmela, 215
Gothein, Eberhard, 110

Greece, *see* Classical circle, culture
Green, Martin, 222
Gundolf, Friedrich, 56–7, 116, 222, 223, 225

Habermas, Jürgen, 18, 26, 214, 216
habitus, 3, 4, 8, 10, 16, 17, 47, 103, 116, 128, 162, 192
Haenisch, Konrad, 101–2, 231
Hauptmann, Gerhard, 32
Hausen, Karin, 218
Hegel, G.W.F., 18, 22, 74, 76, 104, 130, 187, 196–8, 241
Heidelberg, 2, 5, 8, 45, 48–57, 61, 82, 100, 103, 106–8, 112–13, 115, 133, 146, 149, 157, 159, 162, 165, 232, 238
 Myth of, 48, 54, 160–2
 Spirit of, 48, 50, 54, 57, 102, 107, 110–11, 113, 160
 University of, 10, 48–52, 55, 108, 110, 113, 161–2, 217, 240
Heidelberg Community, 221
Heimat, 49
Heller, Hermann, 106–7, 141
Hennis, Wilhelm, 1, 211
Herkner, Heinrich, 222
Heuer, Gottfried, 222
Hildebrand, Bruno, 215
Hildebrand, Johannes, 214
Hintze, Otto, 224
Hirschman, C.F., 163, 234
historians, 20, 111, 117, 122, 130, 167, 168, 170
historical body, 120, 122–4, 129–30, 132, 142, 154, 166, 170, 172, 174–80, 182, 183–5, 188–90, 192–5, 198–203, 239
historical circle 120, 166, 169, 170, 176, 183, 184, 192, 234, 239
Historical School of National Economy, 5–6, 8, 15, 18, 20–9, 31, 38–41, 43, 45, 47, 55, 61, 76, 81, 119, 128, 213, 214, 215, 216, 217
 third generation of, 6, 27–9, 55, 119, 133
historicism, 17, 53, 103, 116, 118
history, 9, 38, 50, 51, 63, 74, 76–7, 79, 106, 118, 120, 124, 128, 130, 131, 147, 149, 158–9, 165–9, 173–4, 176–8, 184, 196–9
 totality of, 168, 171

Hobbes, Thomas, 217
Hoeges, Dirk, 238
Holborn, Hajo, 221
Hölderlin, Friedriich, 50
Honigsheim, Paul, 221
Hübinger, Gangolf, 223
Hughes, H. Stuart, 211, 233, 234
Humboldt, Wilhelm von, 53, 221
hyperinflation, *see* inflation

Idealism, 63, 69, 91, 104, 140, 184, 191, 195
"Ideas of 1914" 58, 60, 138, 156
ideational studies (*Ideen-Studien*), 26–9, 31, 40, 45, 54, 61
Incalcata, 221
India, 117, 131, 166, 169, 172, 176–7, 180–4, 192–3, 202
individual, individualism, 2–6, 10, 21–2, 48–9, 53, 55, 63–4, 69–76, 75, 82, 85, 88, 92, 94, 105, 115, 120–7, 129, 131, 134–6, 138, 141, 144–6, 150, 157, 159, 196–8, 219, 223
individuality, 9, 10, 49, 53, 64, 74, 91, 94–5
industry
 domestic, 32–9, *also see* clothing, ready-made
 factory, 32, 34–7
 location of, 38–45
inflation, 2, 4, 100–1, 143–4, 160
Institute for Social and State Sciences (InSoSta), 110–13, 146, 162, 233
Institute for Social Research, Frankfurt, 110
intellectualization, intellectualism, 56, 74–5, 79, 80, 85, 121, 124, 149–50, 173–5, 180–1, 185–7, 203
intellectuals, intelligentsia, 2, 8, 16–17, 19, 39, 44–5, 47, 50, 52–5, 58, 67–8, 100, 103, 108–10, 112–13, 134, 136–7, 142–4, 146–7, 150–160
 free-floating, 147, 150–1, 239
 "rentier," 143, 146
 "worker," 143–4, 146, 228
intuition, 66, 67, 74, 127, 138, 186–7, 189, 223, 226, 241
iron cage, 55, 134, 136, 223, 226
Isard, Walter, 220

Islam, *see* Mohammedanism
isodapanes, 42

Jacobson, Arthur J., 232
Jaffé, Edgar, 29, 55, 110
Jaffé, Else, 55, 59, 64, 89, 218
Jansen, Christian, 211, 212, 220, 221, 224, 232, 240
Janus Circle, 57
Jaspers, Karl, 161
Jellinek, Georg, 105–7
Jellinek, Walter, 106–7, 111
Jews, 79–80, 110, 228
Junker, 13, 26, 109, 118

Kafka, Franz, 2, 39, 228
Kahler, Erich von, 56, 223
Kalberg, Stephen, 235
Kant, Immanuel, 175, 177, 181, 196
Käsler, Dirk, 217, 231
Kelsen, Hans, 105, 141
Kettler, David, 222, 231
Keyserling, Hermann, 57, 68, 221, 224, 225, 233
Keyserling, Leonie, 225
Knies, Karl, 214
Kocka, Jürgen, 228
Köhler, Walter, 77–8, 227, 228
Kolk, Rainer, 221
Korean War, 160
Kosselleck, Reinhard, 214
Kroeber, A.L., 211
Krüger, Dieter, 216
Kruse, Volker, 16–17, 24, 54, 103, 121, 212, 213, 214, 216, 218, 221, 235, 236
Kühn, Johannes, 240

Laband, Paul, 104–5, 140
Lamprecht, Karl, 197
Landsberger, Adolf, 79
Lange-Kirchheim, Astrid, 218
Laqueur, Walter Z., 221, 224
Lauer, Gerhard, 223
leader democracy (*Führerdemokratie*), 9, 112, 138, 139, 142
leadership, 8–9, 19, 20, 52, 100, 105, 117, 142, 156
  spiritual, 9, 60, 72, 82, 95, 100, 108, 136–9, 141–2, 151–2, 154, 156
  plebiscitary, 107, 136–7, 139

Lederer, Emil, 110, 113, 146–8, 231, 238
Leibniz, Gottfried Wilhelm, 49
Lenger, Friedrich, 228
Lenk, Kurt, 162, 240
Lepennies, Wolf, 213, 223
Lepsius, M. Rainer, 220–1
Lichnowsky, Leonore Gräfin, 219
Lichtblau, Klaus, 216, 221
Lieber, Hans-Joachim, 217
Lietzmann, Hans J., 233
life-aggregation, 122, 124–5, 128–9, 132, 188–90, 192, 195, 203, 236
life-feeling, 74, 75, 77, 78, 80, 123, 190–2, 194–5
life-stream, 120, 123
life-substance, contents, 7, 127, 190, 201
Lindenfeld, David F., 216
Loader, Colin, 212, 222, 223, 227, 228, 231, 238
Lösch, August, 220
Ludendorff, Erich, 92, 99
Ludwig, Emil, 221
Luhmann, Niklas, 4–5, 213
Lukács, Georg, 57, 73, 77–8, 115, 146, 149, 213, 217, 224, 227, 228, 234
  History and Class Consciousness, 115, 149
Luoma, Matti, 130, 236

mandarin, 15, 212
Mannheim, Karl, 52–4, 110, 113, 115, 146–50, 160, 220, 221, 223, 231, 238, 239
  "Competition as a Cultural Phenomenon," 149
  Ideology and Utopia, 147, 150, 239
Marshak, Jakob, 111
Marx, Karl, 1, 74, 139, 145, 149, 196
Marxism, 21, 24, 115, 124, 128, 141, 147–8, 197–8
Masaryk, Thomas, 39, 239
masses, 9–10, 22, 37, 56, 63, 71–2, 86, 108, 118, 136–9, 141, 144, 159, 166
"massification," 101, 107
Masur, Gerhard, 220
Max of Baden, 138
McClelland, Charles, 214
Mediterranean culture, 169
Meinecke, Friedrich, 111
Meinhold, Helmut, 240

Menger, Carl, 219
metahistory, 9, 115, 117, 120, 130,
    also see philosophy of history
metasociology, 159, 160
methodology, 8, 23, 38-9, 77-8,
    115-16, 125-9, 150, 159, 167,
    234, 235
Meyer, Henry Cord, 91, 230
Michels, Robert, 2, 136, 211
Mierendorff, Carlo, 221
Miller, Louis, 214
Mitteleuropa, 9, 82, 89, 91-5, 105, 108,
    112, 136, 152-3, 158, 161, 218, 224
Mitzman, Arthur, 222, 223
Mohammedanism, 182, 191
Mommsen, Theodor, 224
Mommsen, Wolfgang J., 237
monadic organicism, 6, 10, 48-9, 53-4,
    59-61, 63-4, 68, 70, 72, 73, 74,
    78, 80, 82, 88-9, 95, 104, 123-4,
    129-31, 137, 139, 141, 153, 155,
    157-60, 162
money, 4-5, 82, 100-1, 119, 137, 145,
    175, 178, 179
morality, 19, 20, 22, 23, 25, 31, 32, 41,
    53, 68, 101, 215, 236, *also see* values
morphology, 120, 130, 150, 167, 195,
    199-201
Müller, Guido, 233
Müller-Seidel, Walter, 228
Müssiggang, Albert, 215, 216
Mussolini, Benito, 111

nation, 4, 20-2, 25-7, 36-7, 38, 45, 53,
    63, 70, 81-2, 85, 94-5, 100, 104,
    106-7, 118, 131, 134, 138, 153,
    158, 194
    as an organic totality, 5, 21, 23, 32,
        36, 47-8, 49, 61, 83, 85, 91, 112,
        133, 137, 140, 157
National Assembly, 109
national community,
    (*Volksgemeinschaft*), 8, 14, 32, 49,
    58, 89, 99, 116, 138, 156
National Economic Institute, 110
National Socialism (Nazism), 9, 48,
    112-13, 155, 159, 161-2, 240
Natural History-Medical Association, 221
Naumann, Friedrich, 29, 82, 91-4,
    105, 152, 224, 230

negative parliament, 13, 135, 136
Newton, Isaac, 42
Niederhauser, Elizabeth, 219
Nietzsche, Friedrich, 14, 27, 225, 234
Norton, Robert E., 51, 221, 223
notables, *see* elite
Nutziinger, Hans G., 212

Occident, 168-9, 172, 176-80, 183-4,
    199-200, 202
officials, *see* bureaucracy

Pan-German League, 92
Pankoke, Eckart, 216
parliament, 9-10, 13-14, 16-20, 22-3,
    26, 29, 47, 58, 81-3, 94, 100, 104-
    8, 112-13, 119, 133, 135-7, 139,
    141-2, 155-6, 158, 239
    lower house of (*Reichstag*), 13, 26,
        83, 95, 99, 100, 133, 135, 155
Parsons, Talcott, 159, 233
party, 9, 16-17, 19, 22, 24, 26, 29, 45,
    71, 82-5, 87, 91, 100, 105, 108-10,
    118, 133, 135-7, 139, 141-2, 145,
    149, 150-2, 155-6, 160, 229, 239
People's League for Freedom and
    Fatherland, 92-3
Peukert, Detlev J.K., 231
philosophy of history, 18, 25, 50, 62,
    117, 130, 165, 180, 195, 197-8,
    221, *also see* metahistory
Platonism, 184
Plenge, Johann, 198, 241
Poland, 95, 152
politics, 8, 9-10, 16, 23, 26, 54-6, 57,
    60, 61, 79, 80, 82, 91, 93, 95, 104,
    105, 107-9, 113, 122, 134-6, 139,
    141, 151, 156, 162, 234
populism, 14, 26, 99
positivism, 17, 24, 47, 105-6, 124,
    140, 225
Prague, 6, 31, 38, 39, 40, 48, 54, 64, 226
Preuss, Hugo, 104-5
Prinzhorn, Hans, 221
private officials, 84-5
professors, 5, 7, 15, 20, 22-6, 28, 47,
    49, 51-2, 58-60, 84, 100-3, 107-8,
    111-12, 116-17, 143-4, 161-2, 212,
    215, 216, 224, 230, 232, 234, 240,
    also see university

Progressive Party, 99
psychic, *see* soul
public, 7, 15, 19, 25–8, 29, 37, 40, 47,
    50, 102–3, 107–8, 117, 137,
    143–4, 234
  cultivated, 15–17, 19, 25, 28, 47, 52,
    57, 62, 81, 101, 103, 116–17, 144,
    161–2
  disintegrated, fragmented, 26, 29,
    47, 141
  mass, *see* masses
public opinion, 15, 16, 19, 20, 23, 26,
    103, 116, 141, 229
public sphere, 18–20, 25–7, 47, 57, 81,
    141, 142

race, 118, 161
Radbruch, Gustav, 50, 106–7, 111, 141
rationalism, 68, 72, 102, 124, 150,
    197, 226
rationalization, 7, 67–8, 74, 83–5, 115,
    121, 124, 131, 173–4, 198
Reill, Peter Hanns, 222, 225
Reinharz, Jehuda, 228
religion, 67, 69, 79, 80, 83, 115, 128,
    131, 148, 167, 171, 173, 182–5, 187,
    190, 192, 196, 199, 201, 226, 228,
    241, *also see individual religions*
Remy, Steven, 161–2, 240
Renaissance, 63, 131, 184, 191, 193–4
Repp, Kevin, 14, 213, 217
revolution of 1918, 100, 108
revolutions of 1848, 140, 153–4
Ricardo, David, 21
Rickert, Heinrich, 50
Riedel, Manfred, 214
Riehl, Wilhelm Heinrich, 48, 220
Riesterer, Beate, 224
Ringer, Fritz K., 211, 214, 231, 232
Rockefeller Foundation, 233
Romanticism, 63, 104, 140, 225
Roscher, Wilhelm, 214
Russia, 38, 52, 58–9, 90–1, 93–4, 99,
    102, 153, 183, 230, 236

Saint-Simon, Henri de, 196–7
Salin, Edgar, 110, 223, 224, 233
Salz, Arthur, 111
Sauerland, Karol, 220
Scaff, Lawrence, 212, 213, 223, 238

Schelting, Alexander von, 235
Schierra, Pierangelo, 215
Schleicher, Kurt von, 111
Schlink, Bernhard, 232
Schluchter, Wolfgang, 2, 211
Schmitt, Carl, 105, 107, 141
Schmoller, Gustav, 5, 21–3, 26, 29, 31,
    35–6, 38–9, 47, 54, 64, 76, 81–5,
    104, 133–5, 155, 214, 215, 216,
    219, 222, 228
  Jahrbuch of, 21, 29, 31, 77, 86
Scholtz, Gunter, 24–5, 216
Schön, Manfred, 215
Schopenhauer, Arthur, 184
Schroeder, Ralph, 212
Schumpeter, Joseph, 219
Schuster, Helmuth, 24, 26, 40, 103, 216
Schwabe, Klaus, 224
Schwartz, Sanford, 225
science, 14–15, 24, 26, 55–6, 66, 76–7,
    102, 103, 110, 115–16, 131, 136, 151,
    161, 167, 196, 199, 212, 215, 224
  cultural (*Kulturwissenschaft*), 51, 149
  human (*Geisteswissenschaft*), 1, 17,
    18, 23, 25, 52, 63, 226
  natural, 17, 23, 42, 73, 149, 174–5,
    177, 180, 226, 234
  political, 151
  social, 15, 23, 28, 45, 51, 81, 110, 161
  specialized, 15, 17, 23, 25, 51, 69,
    84–5, 91, 116, 117, 131, 137
Sheehan, James J., 216
Simmel, Georg, 1, 49, 52, 57, 220, 225
Simon, Christian, 214
Sinzheimer, Hugo, 106–7, 111, 141
situational studies (*Lage-Studien*),
    26–9, 31, 38, 40, 217
Sixth Congress of German
    Sociologists, 149
Smend, Rudolf, 106–7, 141
Smith, Adam, 21, 24
Smith, Woodruff D., 216
Social Democratic Party of Germany
    (SPD), 26, 99–100, 102, 109, 155–6,
    158, 160, 162
Social Policy, 8, 23, 25, 27–9, 31, 38, 45,
    57, 69, 81–2, 85–7, 108, 137, 139, 227
Social Policy Association, 21–3, 26–7,
    29, 31, 35–6, 38, 54–5, 81, 83, 87,
    143–4, 215

social process, 6, 8, 73, 120–4, 127–8, 131, 166, 168–76, 186, 188–9, 192–3, 195–6, 200, 202
social question, 14, 22, 31–2, 48, 54, 81, 83, 85, 87
socialism, 36
society (*Gesellschaft*), 14–15, 17, 23, 25–8, 32, 52, 56, 58–9, 70–1, 82, 91, 106, 118, 122, 129, 140–1, 144, 146, 148, 162, 189, 192, 194, 197, 201–3, 217, 236, 237, *also see* civil society
Sociological Discussion Evenings, 57, 110, 113, 160
sociology, 1–3, 7, 9, 49, 75, 77–8, 102, 108, 110, 111, 121, 135–6, 148, 150, 158–9, 170, 195, 214
  of culture, 2, 6–9, 39, 43–5, 55, 59, 61–3, 70, 73, 75, 77–9, 82, 88, 93, 95, 102, 110–12, 115, 117, 120, 122, 124–5, 127–30, 143, 147, 154, 158, 160, 165–204, 212, 218, 227, 234
  of knowledge, 3
Sombart, Werner, 24, 28–9, 44, 58, 79, 158, 217, 219, 224, 228
  The Future of the Jews, 79
  The Jews and Modern Capitalism, 79
  Modern Capitalism, 28
soul (*Seele*), psychic (*seelisch*), 17, 21–2, 53, 66, 69, 84, 93–5, 118–19, 121, 123–5, 127, 129–30, 138, 150, 153–4, 158, 179–85, 187–95, 198–204, 235, 241
Spann, Othmar, 219
Spencer, Herbert, 74, 197
Spengler, Oswald, 8–9, 108, 116–19, 122, 124, 130, 160, 224, 233, 234, 241
  The Decline of the West, 116–19
spirit (*Geist*), 9, 17, 19, 22–3, 48, 51, 53, 60, 62, 69, 70, 74, 76, 85–7, 90–1, 93, 103–4, 107–8, 116, 119, 121, 124, 129, 133–4, 141, 143–4, 146–9, 153–4, 158, 167–8, 170, 172–3, 175, 176–9, 181–2, 186–7, 195–8, 202, 215, 226, 235
  of nation, 20, 21, 28, 58, 102, 104, 107, 138, 186, 191
  objective, 69, 187, 237
spiritual-cultural sphere, 122, 170–3

spiritual currents, 149, 167, 170
spiritual sociability, 52, 110
spiritual workers, 109, 143–4
Spranger, Eduard, 53, 222
state, 5–7, 14–15, 17–18, 20, 22–9, 31, 35, 37–8, 47, 58, 61, 63–4, 70, 76, 84–6, 92–3, 100–2, 106–7, 118–19, 131, 133–5, 139–42, 146, 151, 157, 162, 178, 193–4, 196, 216–17, 228
  authoritarian, *see* bureaucratic *below*
  bureaucratic ("constitutional"), 16, 18–19, 20–3, 26, 28, 33, 47, 58, 62–3, 81–3, 85, 88, 100, 104, 113, 137–41, 151, 155, 156–7, 216, 229, *also see* bureaucracy
  of laws (*Rechtsstaat*), 19, 103–5, 106, 140
  parliamentary, 18–19, 26, 108, 145
  power (*Machtstaat*), 91, 140, 154
State Party, 156
Staudinger, Hans, 70–2, 76–8, 104, 124, 227, 228
steel casing, *see* iron cage
Steinberg, Michael Stephen, 221
Stern, Fritz, 213, 217
Stinnes, Hugo, 109
Stirk, Peter M.R., 232
Stölting, Erhart, 212, 216
Stresemann, Gustav, 152
Struve, Walter, 237
sweating-system, *see* clothing, ready-made
Sweeney, Dennis, 213
Sybel, Heinrich von, 20
synthesis, 9, 14, 17, 19, 23, 25, 52–3, 75, 102–3, 116, 132, 141–2, 147, 149–52, 154, 190, 196, 227
system, *see* field

Die Tat, 50, 70, 111
  circle of, 112
Tenbruck, Friedrich H., 163, 211
third man, 131, 159
Thoma, Richard, 106, 107, 111
Thornhill, Chris, 232
Thünen, Johann Heinrich von, 40–1
Thyssen, August, 109
Tompert, Helene, 221

Tönnies, Ferdinand, 7, 24, 28–9, 70, 76, 115–16, 119, 124, 157, 217, 236
Gemeinschaft und Gesellschaft, 28, 76, 115, 217, 233
The Transformation, 160
Treaty of Versailles, 152, 154
Treiber, Hubert, 220
Treitschke, Heinrich von, 91
Troeltsch, Ernst, 53, 103, 116, 183, 221, 222, 231, 233, 241

unions, 87, 92, 105–6, 133, 144
unity, organic, 6, 9, 10, 12, 19, 22, 29, 32, 47–9, 61, 70, 73, 92, 95, 106–7, 125, 129, 133, 136, 139–41, 146, 148–51, 157–8, 160
university, 2, 5, 13–15, 17–18, 20, 22, 24, 28, 47, 101–2, 108

values, 3, 5, 7, 14, 16, 19, 25, 32–3, 47, 55, 56, 58, 71, 76, 81–2, 84, 90, 91, 101–3, 107, 112, 123, 124, 134, 136, 141, 149, 153, 155, 159, 167, 183–4, 197, 224, 228
vital impulse (*élan vital*), 64–5, 88, 129, 156
vitalism (*Lebensphilosophie*), 6, 14, 27, 39, 43, 47, 49, 52–3, 55–7, 59, 61, 64–8, 73, 88, 108, 116–17, 119, 131, 134, 217, 225, 233
vocation (*Beruf*), 56, 62, 82, 86–8, 102, 115, 134, 136, 143–4, 151, 224, 225, 230
Vondung, Klaus, 214

Wagner, Adolf, 216
Wagner, Peter, 3, 15, 212, 214, 216
Wald, Salomon, 235
Weber, Alfred
  The Crisis of the Modern Idea of the State in Europe, 139–42
  Cultural History as Cultural Sociology, 9, 111, 130–2
  "Cultural Problems in the Era of Capitalism," 61
  "Cultural-sociological Attempts: The Ancient Egyptians and Babylonians," 127–8
  "The Cultural Type and Its Transformation," 62

  The Distress of the Spiritual Workers, 101, 143–6
  The End of Democracy, 154–5
  "Fundamentals of Cultural Sociology," 163–204
  "Konstitutionelle oder parlamentarische Regierung in Deutschland?", 83–5
  "Kultursoziologie," 120, 129–30, 234
  "The Official," 7, 28
  "Religion and Culture," 64, 67–9, 73, 78, 80, 139
  "Tasks and Methods," 124
  Theory of the Location of Industries, 6, 38–45, 54, 84, 93, 222
Weber, Marianne, 51, 110, 160, 213, 221, 222, 223, 232
  circle of, 162, 224
Weber, Max, 1–3, 7, 8, 10, 13, 16, 17, 24, 26–9, 37–8, 40, 50, 55–7, 61, 63, 68, 74, 76, 81, 95–6, 102, 105, 109–10, 115, 124–30, 133–9, 146, 148, 150–1, 159, 163, 212, 216, 219, 222, 223, 235, 236, 237, 238, 239, 241
  circle of, 55, 115, 224
  Economy and Society, 115, 126–7
  ideal type, 125–6, 129, 160
  "Parliament and Government in a Reconstructed Germany," 134, 136
  "Politics as a Vocation," 56, 115, 134, 151
  The Protestant Ethic and the Spirit of Capitalism, 28, 40, 115
  "Science as a Vocation," 56, 102, 115, 134, 136, 151, 224, 238
  social action, types of, 126, 135
  understanding, concept of, 125
Wehler, Hans-Ulrich, 163, 213, 241
Weimar Circle, 110–13, 233
Weimar coalition, 100, 113
Weimar constitution, Article 48 of, 156
Weimar culture, 101
Weltner, G.H., 163, 234
Weltpolitik, 59, 92, 152, 154
Werkbund, 57, 224
Whimster, Sam, 222
Widdig, Bernd, 100, 117, 143, 145, 211, 231, 233, 234, 237
Wiese, Leopold von, 9

Wieser, Friedrich von, 39–40, 219
Wilhelm II, 14
will, 9, 21, 22, 34, 51, 71–2, 74, 83, 99, 104–7, 121, 123, 127, 140, 150–2, 154–5, 156, 168, 170, 174, 187, 192, 195, 200, 239
Willi, Victor, 236
Wilson, Woodrow, 94
Winkel, Harald, 214
Wirsing, Giselher, 111
World War I, 6, 8, 17, 57–60, 81, 89–96, 133, 165, 224

working class, 8, 14, 22, 23, 26, 31–6, 40, 44, 48, 49, 69–71, 76, 81, 84–7, 91, 101, 143, 202, 222, 223, 225
Wynecken, Gustav, 69

youth movement, 14, 47, 52, 57, 62, 69, 70, 73, 110, 239, *also see* Free German Youth

Zimmer, Heinrich, 221
Zuckmayer, Carl, 221, 224, 232
Zwiedineck-Südenhorst, Otto von, 228

CPI Antony Rowe
Chippenham, UK
2017-03-09 07:35